A
HISTORICAL ACCOUNT
OF
USEFUL INVENTIONS
AND
SCIENTIFIC DISCOVERIES:

BEING A MANUAL OF INSTRUCTION AND
ENTERTAINMENT.

BY GEORGE GRANT,

AUTHOR OF "PANORAMA OF SCIENCE," "THE
HISTORY OF LONDON,"
ETC. ETC.

PREFACE.

It has been demonstrated that the desire of obtaining knowledge is one of the most natural, and, at the same time, most ennobling attributes of the human mind. There is at the present time a great number of inquiring minds among the working classes of this kingdom, and a still greater number of the young of all classes thirsting for information, who in entering upon a course of general reading must be greatly at a loss for many things which are familiarly alluded to in ordinary conversation, with which everybody is understood to be acquainted, or would have people to think so, but which, in reality, are only familiar to persons who have been living for a considerable time in intimate converse with the world.

The "Historical Account of Useful Inventions and Discoveries in Science," is intended in some measure to supply such information to the anxious inquirer after knowledge. Of the numerous articles here treated of, it will be perceived that each has been traced to its origin in as lucid a style as possible, and in so doing we have endeavoured to combine instruction with amusement. As a proof of this we need only refer to the table of Contents.

USEFUL INVENTIONS.

PRINTING.

AMONG the many arts and sciences cultivated in society, some are only adapted to supply our natural wants, or assist our infirmities; some are mere instruments of luxury, calculated to flatter pride, to gratify vanity, and to satisfy our desires of every description; whilst others tend at once to secure, to accommodate, delight, and give consequence to man. Of this latter kind, Printing undoubtedly stands pre-eminent; and if viewed in its full extent, it may be truly said to possess a very considerable portion not only of the comforts, but the conveniences and positive utilities of life. The advantages derived from this invention must be acknowledged by all,—this art has proved the principal step towards civilization: by it has Christianity been propagated; and by its powerful means are we made acquainted with all that is useful in knowledge, in art, and science. It would take the pen of an inspired writer to enumerate all the blessings which flow from it. It is a patent engine which possesses a preponderating influence over the mind of man either for good or evil, according as it is used.

As we proceed we will have frequent occasion to express our feelings in grateful eulogium, when considering the benefits resulting to society from various ingenious inventions and discoveries; but when we consider the advantages derived from the typographic art, it appears like a vortex, drawing every other sensation into its deep interest, and engulphing every consideration, so that we can think of nothing but printing, and its extensive catalogue of benefits. This interest is wonderfully increased, whether it be viewed on account of its ingenuity, the extent of its benefits, or the benevolence of its objects. In whatever point of view we behold it, whether as a medium for giving the utmost facility to the despatch of the common concerns of life; or as affording the eager mind of the philosophic inquirer the ready means to gratify the inquisitive thirst of his knowledge; in every species of mental intelligence, the rapid facility which it affords to the multiplication of those mediums of communication, by which knowledge is promulgated in every part of the earth. We are at a loss for a term sufficiently comprehensive to express our sense of the infinite importance of those advantages which accrue to mankind from the invention of an art so replete with important consequences, which we hourly perceive to emanate from typography. We need therefore scarcely offer an apology for inserting a brief history of this divine art in our pages.

The earliest specimens of printing which have been discovered, consist in the stamped marks on the bricks and tiles used in building the tower and city of Babel, and which may be dated as far back as two thousand two hundred years before Christ. A number of these stamped clay materials of Babel are still preserved in antiquarian repositories. It is remarkable that they generally differ in shape and appearance, and that the letters or words, which are in ancient character, seem to have been stamped by the hand with moveable blocks. In Trinity College, Cambridge, some curious specimens are preserved, one of which is a round piece of clay, seven inches in height, and three in thickness at the end, resembling a barrel, being thickest at the middle. This interesting relic, this Chaldean book, is entirely covered with lines of letters and words running from the one end to the other; from its portable character it may be called a *pocket volume*, and one which cannot be less than four thousand years old. It is mounted on a marble pedestal, covered with a glass case, secured by an iron bracket, and so contrived that the curious inspector may cause it to revolve on its marble base; but the greatest care is taken of this valuable relic of antiquity. It appears to have been printed by two moulds, and at the middle of the circumference a small blank square has been left, in case as it is supposed, room should be required for a portion of the clay to escape in the action of compression.

Next to these extremely ancient stamped bricks, in point of interest and antiquity, are specimens of the earliest engraving of letters on stone. We are informed by various historical writers that Cadmus, a Phœnician, who lived one thousand five hundred years before Christ, at a period contemporary with Moses, and who was esteemed as the builder of the city of Thebes, was the first who taught the Greeks the use of alphabetic symbols, an art he most likely acquired from the Hebrews. The most ancient specimen of an engraved inscription now known to be extant, is the Sigean Inscription, so called from having been disinterred upon a promontory named Sigeum, situate near the ancient city of Troy, in Phrygia. It is engraved on a pillar of beautifully white marble, nine feet high, two feet broad, and eight inches thick, and which, from the inscription, served as the pedestal of the heathen god Hermocrates. The letters used in this inscription are the capitals of the Grecian language, though rudely cut, but read from right to left like the Hebrew. This specimen of engraving must be about three thousand years old.

Another not less interesting relic of the earliest age of printing is found in a Roman signet ring or stamp, approaching in character to that species of stamp now used by the post-office on letters. This curiosity is preserved in the British Museum. It is the very earliest specimen we possess of printing, by means of ink or any similar substance. It is made of metal, a sort of Roman brass; the ground of which is covered with a green kind of verdigris rust, with which antique medals are usually covered. The letters rise flush up to the elevation of the exterior rim which surrounds it. Its dimensions are, about two inches long, by one inch broad. At the back of it is a small ring for the finger, to promote the convenience of holding it. As no person of the name which is inscribed upon it is mentioned in Roman History, he is therefore supposed to have been a functionary of some Roman officer, or private steward, and who, perhaps, used this stamp to save himself the trouble of writing his name. A stamp somewhat similar, in the Greek character, is in the possession of the Antiquarian Society, of Newcastle-upon-Tyne.

It will be perceived that however curious these relics of antiquity may be, they do not bear any connection with the art of printing books. The origin of this invention seems to be quite independent of a preceding knowledge of impressing by means of stamps. What is, however, worthy of remark, the art of printing books, though on a rude principle, was known and in use among the Chinese, at least one thousand four hundred years before it was invented in Europe. The printing of the Chinese has never resembled anything of the kind in this country. From the first it has been conducted without moveable types. Each page has been, and continues to be, a block or cut stamp, which is thus useful for only one subject—so that every book must have its own blocks. No press is used. The paper being thin, when laid on the block receives the impression by being smoothed over with a brush. There is reason to infer that the art of printing, as thus practised by the Chinese, may have originated through a knowledge of the still more ancient Chaldean mode of printing by blocks on clay. But we may expect, from the well-known ingenuity of the Chinese, and their (in general,) having the organ of imitation so fully developed, that they will not much longer continue this primitive method of printing, as an enterprising practical printer has emigrated, with an excellent assortment of presses, types, &c., from Edinburgh, to conduct his business in the celestial empire. We wish him all success.

The discovery of the art of printing with moveable types, which took place in the fifteenth century, in Germany, was considerably aided by a fashion, which had been some time prevalent, of cutting blocks of wood into pictures, or representations of scenes illustrative of Scriptural history, and printing them on paper, simply by the pressure of the hand, a brush, or cushion behind.

One of the earliest of these wood-cuts is still extant, and represents the creation of man, as detailed in the book of Genesis. In the centre of the picture stands a figure, intended for the Divinity, having the appearance of an old man with flowing garments, a venerable beard, and rays proceeding from the head; on the ground, before him, lies a human being, intended for Adam, fast asleep; and from an opening in his side is seen proceeding the slender figure of a female, meaning Eve, who is taken by the hand by God, and is apparently receiving His blessing. The execution of this, and cuts of a similar nature, is of the rudest description, and is a striking testimony of the low scale of art at the time. Pictures of this nature, which were bound up into books, nevertheless, were the immediate forerunners of the great invention itself. Books of prints, it will naturally be imagined, would soon be found imperfect, for want of descriptive text; this, therefore, urged on the great discovery. The manufacturers of the books, at first, cut single sentences or words, and stamped them below the pictures; but this not conveying a sufficient idea of the subject represented, an anxiety arose to give a lengthened description on the opposite pages. This it seems was, at length, accomplished; still the sentences were all cut in a piece, and the notion of having separate letters, so as to form words at pleasure, was unknown at that period. We will now proceed to the introduction of the modern art of printing.

Ever since the typographic art has been introduced into modern Europe in its present form, the best, and one of the most certain criterions,—which prove the undoubted sense of our species,—exists in the multiplicity of claims which have been made by several cities for the honour of affording the earliest shelter to the infancy of this art. It really appears to be a question yet undecided, to what city, individual, or even era, to attribute this beneficial invention.

However, there is every reason to believe that in this art, as well as in most others, the improvements which have subsequently taken place, have benefited the art itself, as much as that has benefited mankind: therefore, the question of its origin does not appear to us to be of so much importance.

Amidst the claims of various individuals, Mr. Bouzer, in his "Origin of Printing," says, that this honour ought to be adjudged to one of the three cities of Haerlem, Mentz, or Strasburg; of which, in his opinion, the first named city has best established her legitimate right. "But it appears," to use his own words, "that all those cities, in a qualified sense, may claim it, considering the improvements they have made upon each other."

The real and original inventor of the modern art of printing, as at first used, and from whence the improved practice is descended, was one Laurentius, of Haerlem;

who, however, proceeded no further than to cut separate wooden letters. There is every reason to believe that, at first, these wooden forms were made upon the principle of the *forma literarum* of the Romans. This Laurentius, it appears, made his first essay about the year 1430; he died ten years afterwards, having first printed the "Horarium," the "Speculum Belgicum," and two editions of "Donatus."

The individual on whom history most generally places the honour of being the earliest discoverer of the art of printing by means of moveable letters, or types, was John Guttenberg, a citizen of Mayence, or Mentz, who flourished from the year 1436 to 1466, in the reign of Frederick III. of Germany. The ingenious Guttenburg was born at Mayence, in the beginning of the fifteenth century, and removed to Strasburg about the year 1424, or, perhaps rather earlier. Here he became acquainted with the above-named Laurentius, with whom he proceeded to Haerlem, and continued in the employment of Laurentius for some time. However, he returned to Strasburg, where, in 1435, he entered into partnership with Andrew Dritzehan, John Riff, and Andrew Heelman, citizens of Strasburg, binding himself to disclose to them some important secrets, by which they would make their fortunes. The workshop was in the house of Dritzehan, who dying, Guttenberg immediately sent his servant, Lawrence Beildick, to Nicholas, the brother of the deceased, and requested that no person might be admitted into the workshop, lest the secret should be discovered, and the *forms* stolen. But they had already disappeared; and this fraud, as well as the claims of Nicholas Dritzehan, to succeed to his brother's share, produced a law-suit among the surviving partners. Five witnesses were examined; and from the evidence of Guttenburg's servants, it was incontrovertibly proved that Guttenberg was the first that practised the art of printing with moveable types in Strasburg; and that on the death of

Andrew Dritzehan, he had expressly ordered the forms to be broken up, and the types dispersed, lest any one should discover his secret. The words given in his order, which were supported by documentary evidence, were these— "Go, take the component parts of the press, and pull them to pieces; then, no one will understand what they mean." In the same document mention is made of *four forms*, kept together by *two screws*, or *press spindles*, and of letters and pages being cut up and destroyed.

It has been asserted that Guttenberg stole the types from Laurentius, with which he repaired to Strasburg, and commenced business; but of this we can find no corroboration. It has also been said that upon this occasion, Guttenberg stole his own materials, but this is likewise unauthenticated.

The result of this law-suit, which occurred in 1439, was a dissolution of partnership; and Guttenberg, after having exhausted his means in the effort, proceeded, in 1445, to his native city of Mentz, where he resumed his typographic labours.

Being ambitious of making his extraordinary invention known, and of value to himself, but being at the same time deficient in the means, he opened his mind to a wealthy goldsmith and worker in precious metals, named John Fust, or Faust, and prevailed on him to advance large sums of money, in order to make further and more complete trials of the art. Guttenberg, being thus associated with Faust, the first regular printing office was begun, and the business carried on in a style corresponding to the infancy of the art. After many smaller essays in trying the capabilities of a press and moveable types, Guttenberg had the hardihood to attempt an edition of the Bible, which he succeeded in printing complete between the years 1450 and 1455. This celebrated Bible, which was the first important specimen of the art of printing, and which, judging from what it has led to, we should certainly esteem as the most extraordinary and praiseworthy of human productions, was executed with cut metal types, on six hundred and thirty-seven leaves; and, from a copy still in existence in the Royal Library of Berlin, some appear to have been printed on vellum. The work was printed in the Latin language.

The execution of this—the first printed Bible—which has justly conferred undying honours on the illustrious Guttenberg, was most unfortunately, the immediate cause of his ruin. The expenses incident to carrying on a fatiguing and elaborate process of workmanship, for a period of five years, being much more considerable than what were originally contemplated by Faust, he instituted a suit against poor Guttenberg, who, in consequence of the decision against him, was obliged to pay interest, and also a part of the capital that had been advanced. This suit was followed by a dissolution of partnership; and the whole of Guttenberg's materials fell into the hands of John Faust.

Besides the above-mentioned Bible, some other specimens of the work of Guttenberg have been discovered to be in existence. One in particular, which is worthy of notice, was found some years ago, among a bundle of old papers, in the archives of Mayence. It is an almanack for the year 1457, which served as a cover for a register of accounts for that year. This would most likely be printed towards the close of the year 1456, and may, consequently, be deemed the most ancient specimen of typographic printing extant, with a certain date.

Antiquaries and Bibliomaniacs have found considerable difficulty in ascertaining by what process Guttenberg manufactured types; but it appears to be the prevalent opinion, that those which he first used were individually cut by the hand; and being all made as near a height and thickness as possible, they were thus put together in the forms. The cutting of these types must have been a tedious, as well as laborious, occupation. This ingenious man, however, soon discovered the mode of casting his types, by means of moulds; for without this great accessory to the art of printing, he conceived it was next to impossible to carry on his business. The art of type-founding is therefore given to John Guttenberg, in which it would appear he has had no competitor for the honour; but, it is but justice to state that the plan of striking the moulds with punches was a subsequent invention of Peter Schoeffer, his successor, who became partner with Faust, and afterwards his son-in-law.

That Guttenberg was a person of refined taste in the execution of his works, is sufficiently obvious to every person who has had the opportunity of seeing any of them. Adopting a very ancient custom common in the written copies of the Scriptures and the missals of the church, he used a large ornamental letter at the commencement of books and chapters, finely embellished, and surrounded with a variety of figures as in a frame. The initial letter of the first psalm thus forms a splendid specimen of the art of printing in its early progress. It is richly ornamented with foliage, flowers, a bird, and a greyhound, and is still more beautiful from being printed in a pale blue colour, while the embellishments are red, and of a transparent appearance.

What became of Guttenberg immediately after the unsuccessful termination of his law-suit with Faust, is not well known. Like the illustrious discoverer of the great Western Continent, he seems to have retired almost broken-hearted from the service of an ungrateful world, and to have spent most of the remainder of his days in obscurity. It is ascertained, however, that, in 1465, he received an annual pension from the Elector Adolphus, but that he only enjoyed this trifling compensation for his extraordinary invention for a period of three years, and died in February 1468.

John Faust, who as we have seen, obtained the materials of Guttenberg, laid claim to the invention, which has been granted to him by several. Having sufficient capital at his command, he pushed the trade with great advantage to himself. In the Bibles which he printed he frequently omitted the capital and initial letters, leaving them blank for illumination in gold or azure; this was designedly done for the purpose of imposing upon the public printed copies for M.S. transcripts. The report which is in circulation concerning Faust, appears to come in support of this assertion: it being said he was at Paris, and offering a quantity of his Bibles for sale as M.S. The French, considering the number of them, and also remarking the exact similarity and accuracy of them, even to a single point, concluded it was impossible for the most accurate copyist to have transcribed them so correctly. They suspected him of necromancy, and either actually indicted him, or threatened to do, as a magician; and by this means obtained his secret: whence came the origin of the popular story of Dr. Faustus, his dealing with the devil, and tragical death.

In 1462, when Mentz was plundered and disfranchised of its former liberties, printing rapidly spread through a great part of Europe, particularly its artizans in that branch of art, settled at Haerlem, Hamburgh, and other places; from Haerlem it travelled to Rome in 1466, when the Roman character was adopted in 1467, and soon perfected.

In the reign of Henry VI., the Archbishop of Canterbury sent R. Turnour, master of the robes, and W.

Caxton, merchant, to Haerlem, to learn the art. These individuals privately prevailed upon one Corselis, an under workman to come to England: and a printing press was established at Oxford. This appears in a MS. chronicle still preserved; it informs us, that the execution of the concern entrusted to Turnour and Caxton cost 1500 marks; and that printing was established at Oxford before there was any printer or printing presses in France, Italy, or Spain.

The University of Oxford press was soon discovered to be too remote from the seat of government, and too great a distance from the sea, other presses were speedily established at St. Alban's and the Abbey of Westminster.

In 1467, printing was established at Tours, at Reuthlingen, and Venice, in 1469; and it is likely at the same period at Paris, where several of the German printers were invited by the Doctors of the Sorbonne, who established a press in that city.

All important as the art of printing is acknowledged to be, yet three centuries elapsed from the date of the invention before it was perfected in many of its most necessary details. At first the art was kept entirely in the hands of learned men, the greatest scholars often glorying in affixing their names to the works as correctors of the press, and giving names to the various parts of the mechanism of the printing-office, as is testified by the classical technicalities still in use among the workmen. From the great improvement of punching moulds for casting types by Schoeffer, as formerly mentioned, till the invention of italic letters by Aldus Manutius, to whom learning is much indebted, no other improvement of any consequence took place. It does not appear that mechanical ingenuity was at any time directed to the improvement of the presses or any other part of the machinery used in printing, and the consequence was, that till far on in the eighteenth century, the clumsy presses, which were composed of wood and iron, and slow and heavy in working, were allowed to screech on as they had done since the days of Guttenberg, Faust, and Caxton, while the ink continued to be applied by means of two stuffed balls, at a great expense of time and labour.

At length, an almost entire revolution was effected in the printing office, both in the appearance of the workmanship and the mechanism of the presses. About the same period the art of stereotyping was discovered, and developed a completely new feature in the history of printing. One of the chief improvements in typography was, the discarding of the long s, and every description of contraction; while, at the same time, the formation of the letters was executed with more neatness, and greater regularity.

Among the first improvers of the printing press, the most honourable place is due to the Earl of Stanhope, a nobleman who will be long remembered for his mechanical genius; besides applying certain lever powers to the screw and handle of the old wooden press, by which the labour of the workman was diminished, and finer work effected; he constructed a press wholly of iron, which is known by his name.

Since the beginning of the present century, and more especially within the last thirty years, presses wholly of iron, on the nicest scientific principles, have been invented by men of mechanical genius, so as to simplify the process of printing in an extraordinary degree; and the invention of presses composed of cylinders, and wrought by steam, has triumphantly crowned the improvements in this art. The alteration effected by steam power has been as great in the printing business, as in any branch whatever; for example, with the old wooden press, it took a man two days to complete 1000 sheets, (that is, printed on both sides); whereas the London "Times," by means of the steam press completes 24,000 in one hour! Almost every newspaper in the kingdom is printed by cylinder-presses, although some are worked by hand instead of steam; they are also used in other departments of the printing business.

The introduction of steam-presses would have been of comparatively little benefit, if it had not been furthered by another invention of a very simple nature, now of great value to the printer. We here allude to the invention of the roller for applying the ink, instead of the old clumsy and inefficient balls. The roller, which is simply a composition of glue and treacle, cast upon wooden centre-pieces, was invented by a journeyman printer from Edinburgh, about thirty years ago, and was so much appreciated by the trade, as at once to spread over the whole of Europe.

Were it possible to conjure up the spirits of the illustrious Guttenburg and his contemporaries within the office of the London "Times," or some other large printing-office, where everything is conducted with rapidity, quietness, and order, John Faust might well think that the printers of the nineteenth century had actually consummated what he was only accused of in the fifteenth—completed a compact with the devil!

As it would be a waste of time for us to pretend to describe the various processes and materials required in this beautiful art, as we are aware that, without actual observation, no conception can be formed,—this we know from experience, and though we might, like many others, have pretended to give a description, we are perfectly aware that we would have been unintelligible to the majority of our readers, and very deservedly laughed at for our trouble by any practical printer who might happen to read our pages; as far as we have gone, however, in giving a brief historical account of the art of printing, we have no doubt it will be found correct, as have consulted the best authorities.

STEREOTYPE.

STEREOTYPE, as we have mentioned in the former article, was introduced about the middle of last century; and as it is so intimately connected with the art of printing, we could not find a more appropriate place than immediately following that noble art.—Earl Stanhope has been named as the inventor; but for this we have not sufficient authority, and it appears extremely doubtful; as stereotyping appears to have been invented simultaneously, in various parts of England and Scotland, by different persons; still it was upwards of sixty years before it was brought to such perfection as to be applicable for any beneficial purpose.

When properly made known, it was hailed with approbation by those more immediately interested—the printers and publishers: but as experience more fully developed its powers, it was found available only for particular work. For the better understanding of this art, which is comparatively little known, we will give a description of the process, which we are enabled to do by the assistance of an experienced workman.

In *setting* the types, they are lifted from the case, one by one, with the right hand, and built in a small iron form, called a *composing-stick*, held in the left hand of the compositor, who sets line after line till the stick is filled, when he empties it upon a *galley*, and commences again in the same manner, till he has got as much up as will make a page; this page he ties firmly up, and places upon a smooth stone, or cast iron table. In this manner he continues, till he gets as many pages as will make a *form*, which consists of 4, 8, 12, or more pages, as the case may be. If this form is to be worked off at press without stereotyping, the pages are all imposed in one *chass*, and carried to press for working, and when the whole of the impression is off, it is thoroughly washed, and carried back to the compositor for distribution—that is, putting the types in their proper places. When these pages are to be stereotyped, they are *imposed* separately, and carried to the stereotype foundry, where they are examined, and all dirt taken from the face; they are then slightly oiled, and a *moulding-frame* put round each. The frame is filled with liquid plaster of Paris, which is well rubbed into the face of the type to expel the air. As soon as this plaster hardens, it is removed from the page, and shows a complete resemblance of the page from which it is taken. The mould is put into an oven to dry, where it remains till it resembles a piece of pottery; it is then put into an iron pan, in which there is a thin plate of the same metal, called the *floating-plate*; it has also an iron lid, which is firmly screwed down, and the whole is immersed in a pot of molten type-metal, which fills the pan by means of small holes in the corners of the lid. The length of time it remains in the pot depends upon the heat of the metal, but it is generally from ten to fifteen minutes, when it is taken out, and put aside to cool. On opening the pan, nothing is seen but a solid lump of metal, which, when carefully broke round the mould, a thin plate is obtained from the mass, exhibiting a perfect appearance of the page from which the mould was taken.—This is called a stereotype plate, which in general is not above the eighth of an inch thick, and is printed from in the same manner as a page of types. Such is the process of stereotyping, which has become pretty general throughout the trade, but is not much known to the public.

ENGRAVING.

ON WOOD.

AS we have shown in our article on Printing, Wood-engraving was in fashion prior to the invention of printing. We are informed by Albert Durer that Engraving on Wood was invented about the year 1520; he may be a good authority in some matters, but in this he has committed a mistake of nearly one hundred years; seeing that there is at least an impression of one engraving on wood, the representation of the Creation, which was in existence prior to 1430. It was undoubtedly a piece of rough workmanship; but what could be expected at that early period of the art? It has been, however, gradually improving ever since, and it has now attained a point of excellence equal to any of the fine arts, and calls forth the admiration of every lover of the beautiful. It would be invidious to select any of the numerous artists now flourishing—perhaps it would be difficult to make a selection where so many are upon an equality; and we are of opinion they themselves are more willing to accept the public approbation as their reward, than any praise our pen could bestow. All we can do is to recommend our readers to examine for themselves; they have abundant opportunities in the numerous illustrated publications that are daily issued from the press, and bestow that meed of praise upon the respective artists they may deem proper.

The process of engraving on *wood* is diametrically distinct and opposite to that of engraving on *copper* or *steel*; as in the former, the shades are produced by the parts of the work which are made most prominent, and obtrude upon the surface of the substance; whence its chief merit has been regarded in leaving broad and well-proportioned lights. The parts to produce this effect being of necessity excavated, great art and a masterly judgment are necessary to effect this, and at the same time not to weaken the substance, lest it should be injured in the pressure necessary to produce an impression.

The substance usually employed for these engravings is wood of a close grain; on this account box-wood is generally selected. The impressions are obtained from wood-engravings upon exactly the same principle as are the impressions from typography; and they can also be worked off at the same time with the descriptive text. This is a superiority which wood possesses over other engravings, and recommends itself to publishers on account of the immense saving in the expense of a double process in procuring copper-plate illustrations for typographical works, and enables them to keep pace with the ruling passion of this literary era—cheap publications.

ON COPPER.

The art of engraving on copper plates, for impressions, is alleged to have been invented by Peter Schoeffer, one of the early printers, and son in-law of John Faust, about the year 1450. The honour of this invention is also claimed by a Florentine goldsmith of the name of Finguires, who dates his invention in 1540. This artist having used liquid sulphur to take an impression of some chasing and engraving he had made, observed a blackness produced by the sulphur left in the deepest parts of his work, whence he obtained an impression on paper.

But we have no hesitation in giving the preference to Schoeffer, who, we have previously remarked, was of an ingenious turn, and assisted Guttenburg in producing moulds for casting his types; in addition to which, some of the books printed by him are ornamented with head and tail-pieces, with other rude attempts at engraving; and likewise because Schoeffer's claim to the honour was acknowledged before Finguires was born.

Of engraving there are various kinds; that called by connoisseurs, the legitimate mode of engraving, is what is termed the *line* or *stroke* mode. Numerous have been the British artists who have excelled in this style, in affording the means of multiplying our graphical productions.

The next species of engraving we will notice is called the *stipple*, or chalk style,—imitations of chalk drawings. Portraits and historical pieces are executed in this style, which the celebrated Bartolozzi brought to perfection.

The third species we will mention, cannot properly be called engraving; the effect is produced by scraping and rubbing; this kind is called *chiaro obscuro*, or mezzotinto; producing prints which have the effect of Indian ink drawings.

A fourth species of engraving is what is commonly used for landscapes, which produces an effect like a pencil water-colour drawing; which is called *aquatinta*.

In all of these kinds of engravings upon copper the artists find the sulphuric acid, or aquafortis, a most powerful agent. Sometimes, indeed, it is suffered to execute the whole of the process of the graver, especially when it is called an etching.

For the same reasons as those mentioned with regard to wood engravers, we shall abstain from naming any of the very eminent artists now living.

We have already observed the mode of obtaining similar effects from wood and copper, are opposite to each other. The manner in which impressions from wood engravings are obtained, has likewise been noticed; and it remains that we observe the mode by which impressions are obtained from copper-plates. The plate is covered with appropriate ink; the surface is then carefully cleansed, leaving ink only in the excavations or lines in the copper. The plate and paper are passed through a roller press of great power, the roller being covered with a blanket, which presses the paper into all the crevices of the plate, and brings away the ink there deposited.

ON STEEL.

For several years steel has been used in great quantities, instead of copper-plates, by engravers. By this fortunate application of so durable, and it may be added, so economical a material, not only has a new field been discovered admirably suited to yield in perfection the richest and finest graphic productions, which the ingenuity of modern art can accomplish, but to do so through an amazingly numerous series of impressions without perceptible deterioration. The art of engraving on iron or steel for purposes of ornament, and even for printing, in certain cases, is by no means a discovery of modern times; but the substitution of the latter for copper, which has invited the superiority of the British burin to achievements hitherto unattempted by our artists, is entirely a modern practice.

In the year 1810, Mr. Dyer, an American merchant, residing in London, obtained a patent for certain improvements in the construction and method of using plates and presses, &c., the principles of which were communicated to him by a foreigner residing abroad. This foreigner was Mr. Jacob Perkin, an ingenious artist of New England, and whose name has become subsequently so extensively known in this country, in connection with roller-press printing from hardened steel plates. The plates used by Mr. Perkins were, on the average, about five-eights of an inch thick; they were either of steel so tempered as to admit of the operation of the engraver, or, as was more generally the case, of steel decarbonated so as to become very pure soft iron, in which case, after they had received the work on the surface, they were case hardened by cementation.

The decarbonating process was performed by enclosing the plate of cast steel properly shaped, in a cast iron box, or case, filled about the plate to the thickness of about an inch with oxide of iron or rusty iron filings; in this state the box is luted close, and placed on a regular fire, where it is kept at a red heat during from three to twelve days. Generally about nine days is sufficient to decarbonize a plate five-eighths of an inch in thickness; when the engraving or etching has been executed, the plate is superficially converted into steel, by placing it in a box as before, and surrounding it on all sides by a powder made of equal parts of burned bones, and the cinders of burned animal matter, such as old shoes or leather. In this state the box, with its contents, closely luted, must be exposed to a blood-red heat for three hours; after which, it is taken out of the fire, and plunged perpendicularly edgeways into cold water, (which has been previously boiled) to throw off the air. By this means the plate becomes hardened without the danger of warping or cracking. It is then tempered or let down by brightening the under surface of the plate with a bit of stone; after which it is heated by being placed upon a piece of hot iron, or melted lead, until the rubbed portion acquire a pale straw-colour. For this purpose, however, the patentee expressed himself in favour of a bath of oil heated to the temperature of 460 degrees, or thereabouts of Fahrenheit's scale. The plate being cooled in water, and polished on the surface, was ready for use.

A more material peculiarity in Mr. Perkins' invention, and one which does not seem to have been approached by any preceding artist, was the contrivance of what are called *indenting cylinders*. These are rollers of two or three inches in diameter, and made of steel, decarbonized by the process above described, so as to be very soft. In this state they are made to roll backward and forward under a powerful pressure, over the surface of one of the hardened plates, until all the figures, letters, or indentations are communicated, with exquisite precision, in sharp relief upon the cylinder; which, being carefully hardened and tempered, becomes, by this means, fitted to communicate an impression to other plates, by an operation similar to that by which it was originally figured. It will be obvious that one advantage gained by this method must be the entire saving of the labour and expense of re-cutting in every case, on different plates, ornaments, borders, emblematical designs, &c., as these can now be impressed with little trouble on any number of plates, or in any part thereof, by the application of the cylinder. At first sight, the performance of such an operation as the one now alluded to may appear difficult, if not impracticable; and, indeed, many persons on its first announcement were disposed to doubt or deny its possibility altogether. With a proper and powerful apparatus, however, this method of transferring engravings from plates to cylinders, and *vice versa*, is every day performed with facility and success, not only in the production of bank notes, labels, &c., but in works exhibiting very elaborate engravings.

LITHOGRAPHY.

LITHOGRAPHY is the art of printing from stone, which claims for its author Aloys Senelfelder, a native of Munich, in the kingdom of Bavaria. The history of this useful art is recorded by the only person capable of assigning proper and correct motives, and of tracing the various means which were employed to arrive at the desired end, to ultimate success: had all other useful inventions, profitable and elegant arts, had the good fortune which this has happily experienced, we should not have had so much cause to regret deficiencies as we have frequently experienced in the course of our inquiries; then would the various illustrious authors of arts have had justice rendered to them, and still have remained possessed of that glorious immortality so justly the reward of transcendant merit; for the history of this meritorious invention is given by the author himself, thereby securing to it those advantages, which the erudite author of the preface congratulates the public upon, when in his concise epistle he uses that beautiful expression of his countryman, Klopstock, where he says, "Covered with eternal darkness are the great names of inventors."

This work has been translated into English, and published with the following title:—"A complete Course of Lithography, containing clear and explicit Instructions in all the different branches and manners of the Art; accompanied by Illustrative Specimens of Drawings; to which is prefixed a History of Lithography, from its Origin, by Aloys Senefelder, Inventor of the Art of Lithography, or Chymical Printing," &c.

The author of the preface to this work, and friend of the inventor, states that this is an art, whereby the artist, a minister, a man of letters, or a merchant, &c., may multiply his productions at will, without the assistance of a second person.

The author of the above work proceeds to give in detail his motives for the original invention, in which he has not only been strictly circumstantial, but no more so than the curiosity of the public requires, which is always excited in a degree proportioned to the confessed utility of a work, or that demand which its elegance has upon cultivated and delicate feeling. His labours may be said to be divided naturally into two parts, of which division the author has availed himself; first, adducing its history, and secondly, affording the operation of its process.

Its history appears to have arisen with its origin; and both to have originated in the necessities of the author. From whence it appears, that after he had received a scholastic education to qualify him for the jurisprudence of his country, the death of his father, who was a votary of the Thespian art, deprived him of those resources essential to enable him to pursue his intended honourable vocation; he was consequently driven to seek support from the previous acquisitions of his mind. He accordingly devoted his earnest attention to solicit the favours of the dramatic muse as an author. After encountering numberless difficulties, he produced one play, which was published, and sold considerably well. But the honourable independence of his mind induced him to reflect upon the certainty of the large expense, which necessarily attends the practice of an author, who has not liberal patrons in the public or the trade; and the uncertainty of adequate remuneration from the public, for whose amusement they make such large sacrifices of time, ease, property, health, and often life itself. These reflections induced his ardent and ingenious mind to endeavour to avoid the uncertainty of this contingency. He did not possess property to enable him to establish himself as a printer, which was his desire; he was therefore compelled to have recourse to his own ingenuity. He tried various, and at first, unsuccessful experiments, which he ingeniously details; because, he considered, that nearly as much is learned from the failures of an artist, had he always the honesty to publish them, as is gained from his most successful discoveries.

Various were the materials upon which he first essayed to complete his purposes; till, at length, chance directed him to try what could be effected upon stone. For this purpose, he used a species found in Germany, of a beautifully close grained and dense kind, susceptible of receiving a fine polish, called Kellheim stone. Knowing the failures which his countrymen had experienced in endeavouring to fix the ink in this stone for etching, he had recourse to a chemical experiment to obviate this, which succeeded in the following manner:—To four or five parts of water, he added one of rectified vitriol, which instantly produced an effervescence, on being poured upon it; the stone was instantly covered with a coat of gypsum, which to vitriol is impenetrable; this is easily wiped off, and the stone being dried, it is ready for use. The next want he found, was a species of ink, proper to answer the peculiar purposes of the material whereon he had to operate; for which he discovered none so well adapted as the following mixture:—A composition of three parts of wax, with one of yellow soap, is melted over a fire, and mixed with a small portion of lamp-black, dissolved in rain-water. But this is now greatly simplified, as the lithographic printers generally use the same ink as the copperplate printers.

The process of lithography is very simple. The article wanted to be printed is written or drawn upon a piece of transfer paper, which being wet and laid on the stone, and put through the press, the writing or drawing remains on the stone, and any number of impressions may be taken off. Care must be taken, before inking, to come over the stone with a damp sponge, to prevent the ink adhering to the places not wanted, which it would otherwise do.

We understand the Bath and Portland stone is successfully used; but the best yet found in Britain, for the purpose, is what is known by the name of *lias*, raised near Stratford-on-Avon, in Warwickshire; it is a calcareous and partly siliceous stone, and we think not destitute of magnesia, having, when, polished, a very silky and somewhat saponaceous feel.

This art has flourished to a greater extent than we believe the most sanguine expectations of its inventor could have anticipated. Many beautiful specimens of art have been produced equal to the finest copper-plate engravings. It is excellently adapted for drawing of plans, bill-heads, circulars, cards, and many other light articles, which used formerly to be printed by means of letter-press; and on account of the numerous ornaments so easily applied to the lithographic process, the most of these, and similar articles, are principally lithographed, to the detriment, we would conceive, of the letter-press and copper-plate printers.

PAPER.

BEFORE the invention of paper, in ancient times, a great variety of substances have been used for the purpose of recording events, or delineating ideas, of which it becomes our duty to give a somewhat detailed account, to show our readers the numerous advantages they enjoy, in having a material which, like everything in common use, is but little thought of. But let any one suppose himself to be without this necessary article, or the means of communicating his ideas, he would be sensible of the difference.

Rough stones and stakes were used as the first known records of the ancient Phœnicians, remains of which are reported to be still visible; and to confirm this persuasion, certain heaps of stones have been discovered in the environs of Cadiz, which are currently believed to be the remains of those monuments alleged to be made by Hercules, in memory of his famous expedition to the gardens of the Hesperides, for the golden fruit, or as others have it, against Spain. It is also stated, that the usual mode of recording great events, in the north of Asia and Europe, was by placing stones of extraordinary size; in aid of this, we have a great variety of instances.

Since the *scriptural* art has been introduced, or invented, many materials have been, in a variety of ages, and in numerous countries, used for the purpose of recording events to posterity; characters cut upon rocks, upon tables of stone, upon bark, pieces of wood, written upon skins of fish and animals, palm-leaves, besides a great variety of other articles, of which we will only enumerate a few.

There is a Bible still preserved, written on palm-leaves, in the University of Gottingen, containing 5,376 leaves. Another Bible, of the same material, is at Copenhagen. There was also, in Sir Hans Sloane's collection, more than twenty manuscripts, in various languages, on the same material.

The protocols of the Emperors in early times were written upon bark. In the British Museum are many specimens of this substance; also in the grand Duke's gallery at Florence.

To this mode is supposed to have succeeded the practice of painting letters on linen cloth and cotton; what was the difference in the preparation of that material to the one now employed is not ascertained, but it is considered that some preparation was necessary in order to use that substance. There have been frequently found in the chests or cases containing the Egyptian mummies, very neat characters written on linen. Linen being subject to accidents from becoming mouldy, &c., asbestoes cloth had been occasionally used in small quantities.

The accidents to which these species of materials were most of them subject, and linen particularly so, induced man to endeavour to remedy those objects; he accordingly is found to have recourse to the animal creation.

In the convent of Dominican monks at Bologna, are two books of Esdras, written on asses' skins, said to have been written by Esdras himself. The ancient Persians wrote on hides, from which the hair was scraped. The shepherds wrote their songs with thorns upon straps of leather, which they wound round their crooks.

The ancient Welch had a peculiar manner of writing upon small squared oblong pieces of wood, which they called *billets,* which name forms the appellative to numerous of their productions, as the "Billett of the Bard."

The Italian kings, Hugo and Lotharis, gave a grant to the Ambrosian church, at Milan, written on the skin of a fish.

In the Alexandrian Library there were the works of Homer, written in golden letters on the skins of animals. In the reign of the Emperor Baliskus, the head and "Odyssey" of Homer, written in golden letters, on the intestines of beasts, one hundred and twenty feet long, were burned at Constantinople.

In the royal library at Hanover, there is a gold plate, written by an independent prince of Coromandel to George II., three feet long and four inches wide, inlaid on both sides with diamonds.

At last we have arrived at the period for the introduction of the Egyptian papyrus, a kind of rush of large dimensions, growing in the marshes on the banks of the Nile. This plant is described as growing in swamps to the height of fifteen feet; the stalk triangular, of a thickness to be spanned, surrounded near the root by short leaves; stalk naked, has on the top a bush resembling the head with hairs, or long thin straight fibres; root brown.

The Egyptian papyrus was manufactured into paper from very fine pellicles near its pith, separated by a pin or pointed mussel-shell spread on a table in such form as was required, sprinkled with Nile water; on the first layer a second layer was laid crosswise to finish the sheet, then pressed, hung to dry, and afterwards polished with a tooth. The Nile water was very carefully used to prevent spots. Twenty skins were the greatest number which could be procured from one plant. Those nearest the pith made the finest paper. Twenty sheets glued together were called *scapus*, but sometimes *scapi* went to form a *volumen*. This part of the business was executed by the *glutinatoris*, who resembled our bookbinders.

This plant yielded materials for making four sorts of paper.

With respect to other substances for the same purpose, there are many, but as most of these have one generic character, being manufactured from the bark of trees, the detail is not here given, as it might not, perhaps, be generally interesting, especially as nothing new appears in this respect.

With respect to the paper now in use, Dr. Blair says, the first paper-mill (in England, we suppose) was erected at Dartford, in the year 1588, by a German of the name of Spiellman; from which period we may, perhaps, date its manufacture in this country.

It appears, however, that it was known in the East, much earlier; it being observed that most of the ancient manuscripts in Arabic and other Oriental languages, were written upon cotton paper, and it is thought the Saracens first introduced it into Spain.

Anderson, in his "History of Commerce," says, that till the year 1690, there was scarcely any paper made in England, but the coarse brown sort. Paper was previously imported from France, Genoa, and Holland.—However, the improvement of this article in England, in consequence of the French war, produced a saving to this country of £100,000 annually, which had been paid to France for paper alone.

After linen and cotton are so much worn as to be unfit for any other purpose, the several kinds are collected together, and the hard seams and other accumulations, which would require a much longer time to prepare proper for the general mass, than would be consistent with the economy of the whole, those shreds are then separated and thrown away; the different kinds are then collected and kept separate from each other. In such a state of separation they are laid in troughs, which are afterwards filled with water, where they are suffered to remain till a species of fermentation takes place; and the separation of the parts formed by art is not only rendered easy, but also, a division may be made of the most minute parts; the separation is then made by machinery. When properly prepared, a sufficient quantity is placed upon a wire frame, or otherwise one formed of cloth; by mechanical pressure, the moisture is extracted, after which the sheets are hung up separately on lines to dry, in a building properly constructed to admit a free circulation of air.

Manufacturers of paper, originally, could only use white rags to make white paper; but Mr. Campbell, in 1792, discovered a method of discharging any colour from rags, by bleaching with oxi-muriatic acid gas, for which he obtained a patent.

The next considerable improvement which appears to have been made in the manufacture of paper, consists in using felt or woollen cloth in conjunction with the wire cloth formerly used, and now of necessity retained, and other processes too voluminous to be inserted here.

The only remaining circumstance we have to mention is, that in the beginning of the present century there was manufactured, in the vicinity of London, a very good printing paper, made entirely from wheat straw; for which manufacture, the inventor obtained a patent, but he did not succeed, we presume, because it is now discontinued. Considerable quantities of paper is now made from straw in France; but it is of a yellow tinge. Paper made from linen is the best.

PAPER HANGING.

THE desire of man, for the gratification of his natural wants, being soon satisfied, he yet is wanting—those artificial wants which arise in the mind, and are the source of his comforts, because their gratification yields him high delight. Having built him a house, to shelter himself from the exigencies of the weather, to enlarge the sphere of his pleasures, he is desirous to ornament it; and because he cannot, perhaps, construct his house of silver, gold, or costly stones, he endeavours, at least, to have an imitation; and gilding, lacquering, painting, or staining is substituted. This idea, we will presume, to have given origin to every species of decorative ornament in the construction of houses—and among the rest to paper-hanging, which is carried on to a greater extent in this country, than at any former period.

The ancient Greeks, according to Archbishop Potter, constructed not only their arms, but also their houses, occasionally of brass, whilst the Romans frequently gilt theirs; they often covered them with costly casings or veneers, sometimes with precious stones. Since they went to such great cost to ornament the outside of their habitations, we need not wonder that they spared no expense in endeavouring to ornament them within.— Those people, however, who could not procure these extravagancies in reality, thought they would, at least, have the nearest imitation of them; accordingly they had recourse sometimes to veneers of those substances they had seen substantially employed by the rich and luxurious, as well for outside ornament as interior decoration; those who could not afford this, had recourse to pigments and the graphic art; for this purpose, the ingenuity of man was employed to devise various modes of ornament and decoration. Hence arose the various kinds of painting, the fresco, scagliolo, &c., and lastly, came staining of paper in use.

To enumerate the various kinds of this, might be attended with very little benefit, because the principle of all is nearly the same. However, it has been remarked that three kinds are deserving of notice. The first and plainest is that which has on it figures, drawn and painted with one or more colours, consisting only of painted paper. The second contains a woolly stuff, dyed of various tints, and made to adhere to the paper, in certain forms, by a glutinous matter; and the third is a species of paper covered with metallic dust. There are other papers used for hangings, which contain a representation of many kinds of stones, of which we understand there is a large manufactory in Leipsic.

There is also a species of velvet paper—a paper covered with sham plush, or wool dyed and cut short, and made to adhere to the paper by some kind of cement, said to have been the invention of an Englishman, of the name of Jerome Lanyer, in the reign of Charles I., for which he received a patent. In the specification it is stated, that he had found out an art and mystery for affixing wool, silk, and other materials, upon linen, cotton, leather, and other substances, with oil, size, and cements, so as to make them useful and serviceable for hangings and other purposes; which he called Londrindina; and he said it was his own invention, and formerly used within this realm.

However, it appears that this invention of Lanyer was afterwards disputed by a Frenchman of the name of Tierce, who said it was the production of a countryman of his, named Francois, who, he stated, had made such before 1620, and supported his assertion by producing patterns, and the wooden blocks with which it was printed, with the dates inscribed upon them. The son of Francois, it appeared, followed his father's business, at Rouen, for more than fifty years, where he died, in 1748. Some of his workmen are said to have left him, and gone to the Netherlands, Germany, and other places, where they sold their art.

It appears that Nemetz ascribes the invention of wax-cloth hangings, with wool chopped and beat fine, to a Frenchman, named Andran, who, he says, in the beginning of the last century, was an excellent painter in arabesque and grotesque figures, and inspector of the palace of the Luxembourg at Paris, in which he had a manufactory for hangings of that kind. It is also stated that a person of the name of Eccard invented the art of printing, on paper-hangings, gold and silver figures, and that he carried on an extensive manufactory for such works.

It certainly does appear that the Germans cannot claim the privilege of invention here, but were behind their neighbours in this art.

One of the most ingenious of the many new improvements is said to consist in the art of manufacturing paper-hangings by affixing to the substance of the proper metallic dust, commonly called Nuremberg dust, by which it acquires the appearance of various costly metals in a state of fracture, varied with glittering particles of differently formed parts; and receiving the light in every direction, produce certainly a novel effect, which is rich and beautiful, while it is obtained at little expense.

The Nuremberg metallic dust is said to have been the invention of an artist of that city, named John Hautsch, born in 1595, died in 1670; his descendants have continued its preparation to the present time. It is produced from filings of metals of several descriptions washed well in a strong lixivious water, then being placed upon a sheet of copper, are put upon a strong fire, and continually stirred till the colour is altered. Those of tin, by this process, acquire every shade of gold colour, with its metallic lustre; those of copper, different shades of flame colour; those of iron or steel, a blue or violet; of tin and bismuth mixed, a white or bluish white colour. The dust tinged in this manner is then put through a flatting-mill, consisting of two rollers of the hardest steel, like those used by gold and silver wire-drawers; for the greater convenience a funnel is placed over them. French covered paper manufactured from this material is called *papiers avec paillettes*. Its lustre is so durable that it is said to continue unaltered for many years even on the walls of sitting apartments. This metallic dust is an article of commerce, being exported from Germany.

As early as the seventeenth century, the miners of Silesia collected and sold, for various purposes, a material they call *glimmer*, being bright, shining particles of various metals, which those mines produce in great profusion; even the black, we are told, acquires a gold colour by being exposed to a strong heat. This was manufactured by the holy sisters of Reichenstein, into a variety of ornaments; with it they decorated their images, strewing over them a shining kind of *talc*. The silver coloured glimmer had not, however, so great a brilliancy or variety as the Nuremberg metallic dust; for which purposes that article has a decided superiority.

For the various purposes to which these ornaments are to be applied, different adhesive substances should be used; in some cases glue would have the effect, to be first drawn over the substance; in others, a strong varnish, in which wax is dissolved; and for others, various kinds of gums.

Those substances being so covered, the dust may be put in a common pepper-castor, and applied by sifting it over the substance to be so covered. Different figures may be drawn with a pencil, and the box of dust shook over them, as far as the extent of the lines covered with glue; the dust will only fasten so far as it meets with what produces adhesion.

PAINTING.

ITS origin is to be traced up to that known source, from whence most of those arts, which humanise society and lend a polish to life, first had being. Diodorus Siculus speaks of bricks burnt in the fire with various colours, representing the natural appearance of men and animals; which is the first fact upon record. As this occurred during the building of Babylon, it is as remote an original as we are, perhaps, authorised to depend upon; although it is extremely probable it might be traced to an anterior date: which conclusion, though made from inference alone, we are allowed to suppose must have been the case; as a knowledge of the nature of pigments must first have been ascertained before the Chaldean artists could have been informed what colours would fade, or what would withstand the operation of the enamelling process in the intense heat necessary to produce the effect. They must at least have understood the difference between vegetable colours, which are the first presented to the senses, and most probably were the first which were used, and those afforded by the mineral kingdom, which alone were proper for the operation they performed. Therefore, the arts of painting and chemistry, we would presume must have made considerable progress prior to the erection of the tower of Babel.

The next people, who, in point of time as well as of importance, offer themselves to the notice of modern Europeans, are the Egyptians; and their perfection in the use of the various colours which constitute the compound idea we entertain when we think of painting, is well known and appreciated; when we may any day consult our judgment by inspecting those beautiful specimens of their eternal mode of colouring we have in the exhibition on mummy-cases in the British Museum, and other depositories of that species of antique preservation. The third people who excelled in giving a beautiful and tasteful variety to surfaces in colouring and effect, were the Etrurians, a people anciently inhabiting a district of Italy, now known as Tuscany. Of the perfection to which they brought the art we may form an adequate and proper judgment by inspecting those beautiful vases preserved in the Hamiltonian collection in the British Museum, and also in some very curious specimens of ancient painting, procured from the ruins of Herculaneum, collected likewise by Sir William Hamilton.

It cannot be doubted, that most distinct societies of men have, after the gratification of their first wants, and when leisure hours permitted the exercise of their ingenious and inventive faculties, invented a great variety of useful and ornamental arts; therefore, there cannot be a question, but various arts of utility as well as of ornament, have been invented by a great variety of people, who all, agreeably to our prior definitions, are well entitled to the distinct appellation of original inventors; consequently in such a case question must evidently submit to the determination of chronology.

Eudora, the daughter of a potter of Corinth, is presumed to have introduced the art into Greece. The art of painting in Greece is also claimed by Sicyon as the original. Mr. Fuseli has beautifully observed in his first lecture illustrative of the former of these two claimants, that "If ever legend deserved our belief, the amorous tale of the Corinthian maid, who traced the shadow of her departing lover by the secret lamp, appeals to our sympathy to grant it." This invention is becoming doubly interesting in that country, first, because of its elegance and utility; and secondly, because it is ascribed to one of the noblest and most powerful passions, which distinguish the human species, the wonderful effects of which have given to humanity the most exalted and illustrious of actions, which ennoble the character of man—to delicate, refined, and almighty love. Numerous artists in the Grecian school brought the art of painting to great perfection.

The restorer of this delightful art in Europe was Cimabue, a native of Italy, who first studied under some Grecian artists, and furnished some admirable productions in fresco, in several Italian churches about the renovation of the arts in modern Italy; since which time, this purely intellectual art has been successfully cultivated in almost all the countries of Europe, certain masters in all schools of which have been eminent for some peculiar eminence.

An analogy has been drawn by comparison between the fascinating effect of music on the ear, and colour on the eye, wherein it is observed the comparison very nearly approximates; whence the term *harmony*, applied to the former, may correctly, and with singular propriety be used, when speaking of the latter. And also, it is said, for the same reason, and proceeding upon the like analogy, the term *tone* is applicable to both; they are accordingly used indiscriminately. Without questioning their propriety, we give in to our sensations, and as far as our judgment goes, believe they are not improperly introduced into the pictorial art.

It cannot be presumed that we should have the temerity to aspire to the task of giving a full and complete description of every variety which constitutes perfection in the art; for this would be to infer professional ability, equal, or perhaps, superior to what any one individual ever was, or, we may venture to say, ever will be, known to possess. Besides this inference, another must be presumed, because perfection in description must also anticipate the most delicate, refined, and, as termed, classically correct taste; neither to these do we assume the possession of such well-known essentials as are positively necessary to its formation. It is, besides, altogether difficult, as the world acknowledges, to fix a standard to the ideal faculty of taste, and which, we hereby take occasion to notice; therefore we hope to avoid the sin of presumption, and trust that our readers will observe that what we do state is upon good authority, if we have not full confidence in our own experience; but our sin, if sin there be, is rather that of omission than of commission—of saying too little, rather than too much.

STATUARY.

THE origin of Statuary, or what we would term its parent—modelling, is of very great antiquity, as we are authoratively informed by the Grecian historians, whose testimony is supported by Monsieur D'Anville and Major Rennel, two of the most eminent geographers of modern times. From them we learn that three massy statues of gold were erected to ornament the temple of Jupiter Belus. Those were erected by the Chaldeans about two thousand two hundred and thirty years before Christ.

There is also sufficient evidence, that the most eminent and intellectual people, subsequent to the Chaldeans, were the Egyptians.

Every individual, who is in the slightest degree conversant with the history of the arts, knows that the Egyptian artisans had from the earliest periods been in the habit of constructing colossal statues of their numerous deities, and also of their benefactors, raised from gratitude and adulation.

To name only a single instance, the immense colossal statue of Memnon, who perished before the fall of Troy, according to Homer: also Ovid, who speaking of his mother Aurora, says,

> "Nor Troy, nor Hecuba could now bemoan,
> She weeps a sad misfortune now her own;
> Her offspring, Memnon, by Achilles slain,
> She saw extended on the Phrygian plain."

Professor Flaxman has informed us, that this celebrated statue, had it stood upright, would have measured ninety-three feet and a half high; calculating from the dimensions of its ear, which is three feet long. We are informed by Dr. Rees, in his valuable Cyclopedia, that sculpture in marble was not introduced till eight hundred and seventy-three years before Christ. But having said this much for the origin, let us proceed to the art; and we candidly acknowledge that it is from the lectures of that truly distinguished individual, Professsor Flaxman, we are principally indebted for our information.

Sculpture in Greece remained long in a rude state; but we need not wonder at that, when we reflect that art is only an imitation of nature. Hence it follows that man, in a rude state of nature, for want of proper principles to direct his inquiries, and determine his judgment, is continually liable to errors, physical, moral, and religious;—all his productions, of what kind soever, partake of this primitive imbecility.

The early arts of design in Greece resembled those of other barbarous nations, until the successive intellectual and natural, political and civil advantages of this people raised them above the arts of the surrounding nations. The science employed by the Greeks may be traced in anatomy, geometry, mechanics, and perspective. From their earlier authors and coeval monuments, Homer had described the figure with accuracy, but insufficient for general purposes.

OF ANATOMY.—Hippocrates was the first who enumerated the bones, and wrote a compendious account of the principles of the human figure; he described the shoulders, the curves of the ribs, hips and knees; the characters of the arms and legs, in the same simple manner in which they are represented in the basso relievo of the Parthenon, now in the National Gallery of the British Museum.

The ancient artists saw the figure continually exposed in all actions and circumstances, so as to have little occasion for other assistance to perfect their works; and they had also the assistance of casting, drawing, and other subsidiary means. The succeeding ancient anatomists did not describe the human figure more minutely or advantageously for the artist, than had been done by Hippocrates, till the time of Galen, whose external anatomy gave example for that analytical accuracy of arrangement followed by more modern artists. Sculpture, however, profited little from Galen's labours, for the arts of design were in his time in a retrogade motion towards anterior barbarism.

The anatomical researches from Alcmæon of Crotona, a disciple of Pythagoras, to those of Hippocrates and his scholars, assisted Phidias and Praxiteles, their contemporaries and successors, in giving select and appropriate forms of body and limbs to their several divinities, whose characters were fixed by the artists from the rhapsodies of Homer, having then become popular among the Athenians.

Phidias was the first in this reformation. Minerva, under his hand, became young and beautiful, who had before been harsh and elderly; and Jupiter was awful, as when his nod shook the poles, but benignant, as when he smiled on his daughter Venus. Apollo and Bacchus then assumed youthful resemblances of their sire; the first more majestic, the latter more feminine; whilst Mercury, as patron of gymnastic exercises, was represented as more robust than his brother. Hercules became gradually more powerful; and the forms of inferior heroes displayed a nearer resemblance to common nature; from which, both sentiment and beauty can alone be given to imitative art. The near approach of ancient art to nature, considering their high advance to accuracy of imitation, should likewise encourage the modern to imitate the ancient artists. The moderns now also enjoy superior auxiliary assistance from engraving, printed books, &c., which the ancients did not possess.

MECHANISM OF THE HUMAN FRAME.—The human figure with the limbs extended, may be inclined and bounded by the circle and square; the centre of gravity, its change of situation, is susceptible of description, and may be exemplified in rest and motion;—running, striving, leaping, walking, rising, and falling. Those principles of motion may be exhibited in a skeleton, by the bending of the backbone backwards and forwards, whilst the limbs uniformly describe sections of circles in their motions, constantly moving on their axis.

DIMENSIONS OF THE HUMAN FIGURE, as exhibited in Grecian Statuary.—The height, eight heads (or usually ten faces); two heads across the shoulders; one head and a half across the hips; three noses, the thickest part of the thigh; two, to the calf of the leg; one, the narrowest part of the shin, &c. The above is the general proportion of the male figure. The female figure is narrower across the shoulders, and wider across the hips than the male.

The *beauty* of the human figure is found in its proportion, symmetry, and expression; it really appears that the beauty of the human figure is the chief or ultimate of beauty observed in the visible works of creative Omnipotence. From thence every other species of beauty graduates in just *ratios* of proportion. From considering the intellectual faculties of man, we assimilate the idea, and connect beauty with utility, as this union of his physical and mental powers unquestionably renders him one of the most beautiful objects in the creation. This consideration leads us involuntarily to a train of thought, suggested by a principle laid down by Plato, "That nothing is beautiful which is not truly good;" which also induces the following corollary, and which is confirmed by reason, and sanctioned by revelation, that *perfection of human beauty consists of the most virtuous soul in the most healthy and perfect body.*

Inasmuch as painters and sculptors adhered to those principles in their work, they assisted to enforce a popular impression of divine attributes and perfections, even in ages of gross idolatry.

In the highest order of divinities, the energy of intellect was represented above the material accidents of passion and decay.

The statues of the Saturnian family, Jupiter, Neptune, and Pluto, were the most sublime and mighty of the superior divinities. Apollo, Bacchus, and Mercury, were youthful resemblances of the Saturnian family, in energetic, delicate, and more athletic beauty: Apollo-Belvidere supplies Homer's description to the sight; he looks indignant, his hair is agitated; he steps forward in the discharge of his shafts; his arrows are hanging on his shoulder.

A youthful and infantine beauty of the highest class distinguish the Cupid of Praxiteles, and the group of Ganymede and the Eagle. The order of heroes or demigods excel in strength, activity, and beauty; Achilles, Ajax, Hæmon, Zethos, and Amphion, are examples in Grecian statuary to establish this remark.

The Giants are human to the waist; their figures terminate in serpentine tails. Ocean and the great Rivers have Herculean forms, and faintly resemble the Saturnian family, and have reclining positions. The Tritons resemble the Fauns in the head, and upper features, with finny tails, and gills on their jaws; their lower parts terminate in the tails of fish.

In the highest class of female characters, the beauty of Juno, is imperious; that of Minerva, wise, as she presides over peaceful arts; or warlike, as the protectress of cities. Venus is the example and patroness of milder beauty and the softer arts of reciprocal communication; of which the Venus Praxiteles and Venus de Medicis are instances. The Greeks had also a Venus Urania, the goddess of hymenial rites and the celestial virtues.

The Graces are three youthful, lovely sisters embracing: they represent the tender affections, as their name implies; while their character gives the epithet *graceful* to undulatory and easy motion. The universe was peopled by genii, good and evil demons, which comprehends every species and gradation from the most sublime and beautiful in Jupiter and Venus, to the most gross in the Satyr, resembling a goat, and in the terrific Pan.

As the public have now an opportunity of consulting many of the objects above referred to, in our great national gallery in the British Museum, those of our readers who can obtain this advantage will do well to pay a visit to that celebrated depository for the relics of antiquity, where they will have it in their power to convince themselves of the truth of the foregoing remarks.

The progeny of Ham, the son of Noah, we find, peopled Egypt, Medea, Chaldea, Phœnicia, and several other adjoining countries. It will be remembered that two of the three sons of Noah possessed these countries which the folly of idolatry overflowed; whilst it was in the line of Shem alone, that the true faith was continued. The Mosaiac narrative is chiefly descriptive of events which occurred in the posterity of that patriarch, because from it the righteous line of the faithful in Abraham, David, Solomon, and ultimately Christ, proceeded. Thus more than two-thirds of the inhabitants of the world were gross idolators: we often find the Omniscience of the Highest forewarning the sacred line to avoid its fascinations. Nay, when, upon more occasions than one, the descendants of the faithful forgot themselves, and those admonitions of the Creator were neglected, we find the sacred race flying before the face of puny foes, which defeat was declared to be from their having prostrated themselves before strange gods: they were bowed thus low in battle. Not to mention their disobedience immediately beneath Mount Sinai, which protracted their journey through the wilderness to forty years, which, perhaps, under other circumstances, would not have required as many days. All those troubles, their subsequent captivities, and national afflictions, were the produce of disobedience. This is one of those means which retributive justice resorts to punish wilful sin; so, however, it was with the seed of Abraham. And so it is presumed to be with the present race of men; either immediate or remote punishment vindicates the Omnipotence of Heaven. From the frequent maledictions we discover in the sacred volume against idol worship, we cannot doubt that it was peculiarly offensive to the Deity. that the great majority of the world were addicted to this proscribed practice is equally certain. And as the Spirit of Truth had declared in the decalogue, that "It would not be worshipped under any form in the heavens above, in the earth below, or in the waters under the earth;" so was image-worship, and consequently the construction of such things, forbidden.

We discover that as this mania infected all nations, tongues, and people, so did not the Israelites escape it; but immediately after their departure from Egypt we find an exact similitude of the sacred calf of the Egyptians, cast in melted gold, which they constructed below Mount Sinai. In Egypt, metallic statues, as well as those of stone, must have existed anterior to that event, as they actually had done to our own knowledge, and long before idolatry had made its appearance in Egypt, it had existed in Chaldea, as already shown.

As that worship had first its being in Chaldea, so had the art of statuary its origin in that country; it was improved, perhaps, in Egypt, and perfected in Greece, from the time of Pericles to that of Alexander, commonly called the Great.

DRAWING.
THE HUMAN FIGURE.

FROM what has been said in the previous article, it would appear that drawing of the human figure was nearly coeval with the art of statuary, or perhaps prior to it in Greece. As there is ample room to suppose the rude aboriginal inhabitants of Greece borrowed their art, as they did their religious and civil policy, from the Egyptians, and in fact from every nation where they discovered anything worthy their attention, so must we suppose they had also this art, in its infancy it is true, from the same people. Upon reflecting for a single moment, we are fully satisfied that the origin of the art now under contemplation came from Egypt. An ancient philosopher expressed himself with great truth, when he said, "Necessity was man's first instructor." We accordingly perceive the necessity of the earliest inhabitants of Egypt to exercise the art of drawing, they having determined to record their transactions by hieroglyphical representation. We have not the slightest doubt but we have now in the British Museum some of the earliest specimens of Egyptian hieroglyphical delineation, in the *sarcophagi*; from its inscription, it has been discovered that that identical monument cannot be less than three thousand five hundred and ninety-eight years old!

Previous to this, we can have no doubt that the art of drawing must have existed.

Like its sister art, sculpture, it received every improvement of which it was susceptible, from the mature conceptions and the delicate hand of Grecian artisans; words are, perhaps, inadequate to convey this art to a second person. Years of incessant labour, with an attention to principles established and found to correspond correctly with nature, are the only means to obtain a just knowledge of its principles, and to judge tastefully of its correct execution.

However, in addition to the rules laid down in the preceding article, we add the following, which have been approved by Sir Joshua Reynolds, by no means a contemptible judge of the art:—

1. That from the crown of the head to the forehead is the third part of a face.

2. The face begins at the root of the lowest hairs that grow on the forehead, and ends at the bottom of the chin.

3. The face is divided into three proportionate parts; the first contains the forehead or brow; the second, the nose; and the third, the mouth and chin.

4. From the chin to the pit between the collar-bones, is two lengths of a nose.

5. From the pit between the collar-bones to the bottom of the breast, one face.

6. From the bottom of the breast to the navel, one face.

7. From the navel to the genitories, one face.

8. From the genitories to the upper part of the knee, two faces.

9. The knee contains half a face.

10. From the lower part of the knee to the ancle, two faces.

11. From the ancle to the sole of the foot, half a face.

12. A man with his arms extended, is from his longest finger on each hand, as broad as he is long.

13. From one side of the breast to the other, two faces.

14. The bone of the arm called *humerus*, i.e., from the shoulder to the elbow joint, is the length of two faces.

15. From the end of the elbow to the joint of the little finger, the bone called *cubitus*, with a part of the hand, is also two faces.

16. From the box of the shoulder-blade, to the pit between the collar-bones, one face.

17. To be satisfied in measures of breadth. From the extremity of one finger to the other, so that his breadth should be equal to the length, it should be observed, that the bones of the elbows with the *humerus*, and the *humerus* with the shoulder-blade, or *scapula*, bear the proportions of a face when the arms are extended.

18. The sole of the foot is one-sixth part of the length of the entire figure.

19. The hand is the length of a face.

20. The thumb contains a nose in length.

21. The inside of the arm, from the place where the muscle disappears, which is connected with the breast (called the pectoral muscle,) to the middle of the arm, four noses long.

22. From the middle of the arm, at the top, to the beginning of the head, five noses.

23. The longest toe is one nose.

24. The outermost parts of the paps, and the pit between the collar-bones of a female, form an equilateral triangle.

The knowledge of the preceding proportions, are as mere rudiments essential to the delineation of the human figure; but they relate to a body in a quiescent state only. The more difficult task remains to become thoroughly acquainted with its actions. To obtain this, a rudimental and even an intimate acquaintance with the skeleton, and assiduous and incessant practice are necessary.

However, the lectures delivered to the Royal Academy have furnished us with the probable extent to which the motions of the human frame may be carried.

First, premising that the motions of the head and trunk of the body are limited by the several joints of the spine.

2. The motion of the body upon the lower limbs takes place at the hip-joints, at the knees, and at the ancles.

3. Those limbs, called great limbs (the whole frame being technically divided, and denominated the upper and lower extremities), have rotatory motions at their junctions with the trunk, by means of a ball and socket joints, at the shoulders and the hips. The analogy of parts between the upper and lower extremities is not carried through the structure of those limbs in the body.

4. The fulcrum of the upper limb is itself moveable upon the trunk, as appears from the extensive motions of the scapula, which so generally accompany the rotation of the shoulder, and supply the limb with a great variety of motion, much more than the lower limb possesses.

5. The junction of the thigh with the mass without motion, called the *pelvis*, limits its rotation to the ball and socket-joint without farther extension.

6. The rotation of the head and neck takes place at the joint between the first and second vertebræ.

7. When the nose is parallel with the *sternum*, the face may be turned towards either shoulder, through an angle of

60 deg. on each side; the whole range of its motion being 120 degrees.

8. The lateral bending of the neck is equally divided between the seven vertebræ; but the bowing of the head, and violently tossing it backward, are chiefly effected at the joint of the skull, and the first bone of the vertebral column called the atlas.

9. Although the preceding motions are consistent with an erect stature of the neck, yet the lateral motions demand a curvature of its whole mass.

10. The movements of the trunk are regulated by rotary and lateral motions, nearly equally divided among the several joints of the vertebræ of the back and loins.

11. The joints or the dorsal or back vertebræ are, notwithstanding, more close and compact than those of the loins; allowing of a wider range for bending and turning in the loins than the back.

12. The sternum and ribs move upward, to assist the chest in the expansion required for respiration; drawing the clavicles and the shoulders upwards in full inspiration, and tend to a contrary motion on expiration. Such movements also, characterise strong action and certain passion, and very apparent in a naked figure.

13. In stooping to touch the ground, the thigh-bone forms an angle of somewhere about 55 degrees with the average direction of the vertebræ.

14. The leg bends upon the thigh at an angle of about 75 degrees, and the line of the *tibia* forms, with the sole of the foot, when that is elevated, an angle of 65 degrees.

15. The whole of this limb is susceptible of motion at the hip-joint forwards to a right angle with its perpendicular position; and backwards to an angle of 20

degrees. The leg will then continue to move by itself to its own angle of 75 degrees with the thigh. Its extreme motion does not exceed 45 degrees.

16. When the shoulders are quiescent, the clavicles usually meet in an angle of 110 degrees at the sternum.

17. The utmost elevation of the upper joint of the arm generally forms an angle of 155 degrees with the vertebræ, and about 125 degrees with the line of its clavicle. The flexion of the fore-arm upon its upper part is confined to an angle of nearly 40 degrees.

18. The whole arm is capable of moving forward or outward through nearly 80 degrees, and backward to nearly the same angle with its perpendicular station.

19. The actions of pronation and supination in the hand, range through all intermediate degrees from a horizontal or perpendicular direction to 270 degrees; but 90 degrees of its rotary motion in pronation comes from the shoulder joint.

20. The palm of the hand admits of flexion and extension to 65 degrees in each direction; its lateral motions are 35 outward, and 30 inward. The flexion of the fingers at each phalanx is a right angle.

But it must be observed that in drawing the joints, very considerable difference is found in their length, from inequality of action. The elbow joint, when bent inward, lengthens the arm nearly one eighth; the same general law operates on the knees, fingers, &c. When a man is at rest, and standing on both feet, a line drawn perpendicularly between the clavicles will fall central between his feet. Should he stand on one foot, it falls upon the heel of that foot which supports his weight.

If he raises one arm, it will throw as much of his body on the other side as nature requires to support the equilibrium. One of his legs thrown back brings the breast forward, to preserve the gravity of the figure: the same will be observed in all other motions of the parts to keep the central gravitation in its proper place.

The equipoise of a figure is of two sorts: simple, when its action relates to itself; and compound, when it refers to a second object.

The equilibrium of nature is constantly preserved; for in walking, leaping, running, &c., similar precautions are taken. By the flexibility of our bodies in striking, according to the proportionate force meant to be employed, the body is first drawn back, then the limb propelled forward, bringing with it the weight of the body.

In striking, lifting, throwing, &c., a greater proportion of force is employed than may be necessary to effect the intended purpose. This is mentioned because, in representation, the force employed in an action should be marked in the muscle producing that action; if it be marked rather stronger than may be necessary, the cause is obvious, for Nature so employs her powers.

In studying this art, students should have selected for them the best examples to copy from at first; then they should draw from the figure as soon as possible, and if it be possible from the best specimens of the antique. Their first drawings are recommended to be made with chalk, and in large proportion; attention to these will communicate ease and freedom to their future performances.

It will be likewise found necessary for them to draw upon geometrical principles; this communicates a truth, which greatly adds to their certainty and confidence, and ultimately to their ease.

This is mentioned, because it will be found that there is no portion of the human frame, quiescent, or in an active state, but what is susceptible of geometrical definition.

Experience and exercise communicate truths which produce certainty, whence come ease and grace.

ARCHITECTURE.

THIS is a science most beneficial to humanity. It is very evident that it must have an extremely ancient origin. The origin of this art is presumed to have been imitated by man, from those natural caves and recesses, which are discovered in various parts of the earth. For in those, it is reported, the first men took shelter from the inclemency of elemental strife, and to avoid the piercing contingencies of ultimate and precarious uncertainty. The oldest buildings in the world are accordingly said to be beneath the surface of the earth; among which are reckoned the famous temple of Elephanta, in the Delta of the Ganges; the Catacombs, in Egypt; and upon the surface of the earth, the tower of Belus, at Babylon; the Egyptian Mausoleum, and the Druidical Temples in Gaul and Britain.

Architecture may well be denominated one of those arts which accommodate, delight, and give consequence to the human species; while at the same time it is calculated to flatter pride, and gratify vanity. If viewed in its full extent, it may be truly said to possess a very considerable portion, not only of the comforts, but the conveniences, the positive utilities, and many of the luxuries of life. The advantages derived from *houses* only are very great, being the first step towards civilization, having great influence both on the body and mind of man. Secluded from each other in woods, caves, and wretched huts, the inhabitants of such recesses are generally found to be men, indolent, dull, inactive, and abject; their faculties benumbed, their views limited to the gratification of their individual and most pressing wants. But when societies are formed, and commodious dwellings provided, where well sheltered, they may breathe a temperate air, amid summer's scorching heat, and winter's biting cold; sleep, when Nature requires, in ease and security; study unmolested; converse and taste the sweets of social enjoyments;—they are spirited, active, ingenious, and enterprising, vigorous in body, and active in mind. If benefits like these previously enumerated result from any art, then will that of the architect claim a decided pre-eminence. When we reflect on the almost infinity of useful purposes to which this art is conducive;—that it erects us temples for the worship of our Creator, the benevolent dispenser of all good things, that it provides us with habitations, where ease and simplicity are agreeably combined;—that it is conducive to our safety, comfort, and convenience, in uniting different districts of the country by the facility of bridges, roads, &c., is contributive to the gratification of our natural wants, and to our safety.

As inhabitants of a great commercial country, the benefits we derive from *naval* architecture are unspeakably great; when we reflect that it operates as a medium of communication between us, an insulated people, and the whole earth, in its remotest colonies; that it serves to convey between our people and the most distant nations the native produce of the respective countries, with the effects of mutual industry; that it clothes, feeds, and furnishes employment to thousands of our fellow-countrymen; and, in a national point of view, our wooden bulwarks have been the wonder of the world, and continues to afford us protection from our enemies, should all other hopes fail. What can surpass its utility in the latter point of view? what can exceed the assistance derived from it? By the criterions formerly mentioned let us determine. We shall find, that of all the arts the world has ever boasted, there are but few, if any, that can claim a superiority.

There are no other designs, whether necessary or superfluous, so certainly productive of their desired object, so beneficial in consequences, or so permanent in their effects, as is the art of the architect. Most other inventions which afford pleasure and satisfaction soon decay; their fashion fluctuates—their value is lost; but the productions of architecture command general attention, and are lasting monuments, beyond the reach of ephemeral modes: they proclaim to distant ages the consequence, genius, virtues, achievements, and munificence of those they commemorate to the latest posterity. The most obvious and immediate advantages of building are, the employment of numerous ingenious artificers, industrious workmen, and labourers of all kinds; converting materials of small value into the most noble productions, beautifying countries, multiplying the comforts and conveniences of life.

But not the least desirable effects of the architect's art, perhaps, remain yet to be noticed, in affording to the numerous train of arts and manufactures, concerned to furnish and adorn the works of architecture, which employ thousands, constituting many valuable branches of commerce. Also from that certain concourse of strangers to every country celebrated for stately structures, who extend your fame into other countries, where otherwise, it would never have been heard of; adopt your fashions, give reputation, and create a demand for your productions in foreign parts; these are circumstances which certainly should not be too lightly valued, and these circumstances result from architecture.

At this day, the ruins of ancient Rome support the splendour of the modern city, by inviting travellers, who flock, from all nations, to witness those majestic remains of former grandeur. The same may be said of many other countries famous for architectural remains. Thus architecture, by supplying men with commodious habitations, procures that health of body and energy of mind, which facilitates the invention of arts: when by the exertion of their skill and industry, productions multiply beyond domestic demand, she furnishes the means of transporting them to foreign markets: whenever by commerce they acquire wealth, she points the way to employ their riches rationally, nobly, and benevolently, in methods honourable and useful to themselves, and beneficial to posterity, which add splendour to the state, and yield benefit to their descendants. She further teaches them to defend her possessions, to secure their liberties and lives from attempts of lawless violence or unrestrained ambition. So variously conducive to human happiness is this art, to the wealth and safety of nations,—so, naturally, does it demand that protection and encouragement which has ever been yielded it in all well governed states.

The perfection of virtuous other arts we have beheld to be a consequence of this; for when building is encouraged, painting, sculpture, and every species of decorative art will flourish of course. It should not, however, be imagined that the heaping of stone upon stone can be of consequence, or reflect honour on individuals or nations. The practice of architecture infers actual art to be an essential preliminary; without this, and having some laudable end in view, it is apt to raise disgust. This art is generally classed under three distinct heads, viz., Civil, Military, and Naval Architecture.

In the first attempts of architecture it was extremely rude, as might naturally be expected. It has, however, from time to time, as improvements have advanced, been raised to relative importance, as the education of the people progressed; and it certainly gives the best record of the mental progress of every people which can be collected. It has always been found to flourish best in free states, and when the rulers have possessed genius, virtue, and good taste. The most eminent era of Grecian architecture was when the Athenian republic was under the direction of Pericles; at this period, also, existed the first of statuaries, Praxiteles. Where that eminent artist and their admirable architects were employed, in the words of Pausanius, "they rendered the whole of Acropolis as an entire ornament." There are various characteristic distinctions to be made in the several orders of architecture which distinguish the Grecian people. The Doric is eminent for primeval simplicity; the Tuscan embraces more ornament; the Ionic unites simplicity and elegance; but the sum of all excellence appears to be united in the Corinthian. The Composite is also a most elegant order, but appears to have added but little to the Corinthian elegance and majesty. Various nations have a great diversity of architecture; as the Egyptian, Persian (distinguished by human figures supporting entablatures), Hindostanee, Arabasque or Marisquo, which are very peculiar, generally having the walls to project most at the top, which is indicative of the natural jealousy of all oriental people; they all regarding their *women* as their chief treasure, it appears meant for their especial protection.

A greater simplicity does not appear anywhere than in the architecture of the Druids, consisting of most extensive circles of immense stones, chiefly raised perpendicularly, with occasionally a larger stone placed upon the apex of two others horizontally.—There are in Great Britain numerous remains of these constructions: the chief are Stonehenge, near Salisbury; at Avebury, also in Wiltshire; Pomonca, in the Orkneys; Rollright, in Oxfordshire. But the most eminent spot for Druid temples was Mona, in Anglesea, in Wales. The reason for such apparently unmeaning erections will be found in their peculiar belief, in the religion they professed.

The Saxon is a very heavy order of architecture. It was used in this kingdom much in the erection of religious edifices, and is frequently found mixed with the Norman in such structures. The grand and most obvious distinction is a semi-circular arch, with massy columns, variously ornamented, and most frequently the columns which support the same arch are diversely sculptured. The chief sculptures of this kind in Britain, are Gloucester Cathedral; Malmesbury Abbey, Wilts; Sedbury Church, Herefordshire; several churches in Rutland, Lincoln, Somerset, Devon, and other counties.

There appears to us to be no order of architecture better calculated for the purpose to which it is generally adopted, than the chaste and pure Norman style, barbarously denominated Gothic. It affords a great variety of light, airy beauty, and tasteful grandeur.

In this country, the Norman order succeeded the Saxon, and we lost nothing by the exchange; for even now, that we have the entire benefit of a choice of the purest Grecian (since its revival by Inigo Jones), it is a matter of taste to be certain; but in our estimation, the chaste Norman is to be preferred to the purest Grecian, for the purposes for which it is intended; and if the means answer the ultimate end, we submit this to be the proper criterion for preference. We find it usually employed in religious edifices; it is pure, light, airy, and cheerful: and we are of opinion that the service of gratitude and thanks to the Creator demands a disposition of mind which these feelings are best calculated to inspire.

Domestic architecture is various, and chiefly regulated by the various purposes for which it is designed. Its characteristic is utility.

CHAIN-BRIDGES.

(See Frontispiece.)

It appears, from a description of bridges of suspension, communicated by R. Stephenson, civil engineer, some time ago, to the "Philosophical Journal," that the first chain-bridge constructed in this country is believed to be one over the Tees, forming a communication between the counties of Durham and York. It is supposed, on good authority, to have been erected about 1741, and is described in the "Antiquities of Durham" as "a bridge suspended on iron chains, stretched from rock to rock, over a chasm nearly sixty feet deep, for the passage of travellers, particularly miners. This bridge is seventy feet in length, and little more than two feet broad, with a hand-rail on one side, and planked in such a manner that the traveller experiences all the tremulous motion of the chain, and sees himself suspended over a roaring gulf, on an agitated and restless gangway, to which few strangers dare trust themselves." In 1816–17, two or three bridges of iron were constructed; the first, by Mr. Lees, an extensive woollen manufacturer, at Galashiels, in Scotland. This experiment, although made with slender wire, and necessarily imperfect in its construction, deserves to be noticed, as affording a practical example of the tenacity of iron so applied.—These wire bridges were suspended not upon the catemarian principle so successfully adopted in the larger works subsequently undertaken, but by means of diagonal braces, radiating from their points of suspension on either side towards the centre of the roadway. The unfortunate fabric next mentioned was constructed on this defective principle. Among the earliest practical exhibitions of this novel architecture in the United Kingdom, may be mentioned the uncommonly elegant and light chain-bridge which was thrown over the Tweed at Dryburgh, in 1817, by the Earl of Buchan, for the accommodation of foot passengers. Its length, between the points of suspension, was two hundred and sixty-one feet, being considered the greatest span of any bridge in the kingdom. This useful structure, the theme of such just applause, and which harmonised so finely with the far-famed scenery of Dryburgh Abbey, was entirely destroyed by a tremendous gale of wind, at the beginning of the year following its erection.—This bridge was subsequently restored upon a more secure system.

CLOCKS.

THE invention of clocks, such as are now in use, is ascribed to Pacificus, Archdeacon of Verona, who died in 846; but they were not known in England before the year 1368. They were ultimately improved by the application of pendulums, in 1657, by Huygens, a Dutch astronomer and mathematician. Although Dr. Beckmann differs in some slight degree from the previous relation concerning clocks, yet he says, "It is sufficiently apparent that clocks, moved by wheels and weights, began certainly to be used in the monasteries of Europe, about the eleventh century." He does not think, however, that Europe has a claim to the honour of the invention, but that it is rather to be ascribed to the Saracens; this conjecture, he confesses, is chiefly supported by what Trithemius tells us, of one which was sent by the Sultan of Egypt to Frederick II., in 1232. He thinks that the writers of that century speak of clocks as though they had been then well known; he adds, that in the fourteenth century, mention is made of the machine of Richard de Wallingford, which has hitherto been considered as the oldest clock known. The fabricator of this machine called it *Albion*.

It appears that clocks had been hitherto shut up in monasteries and other religious houses, and that it was not till after this time they were employed for more general purposes, as the convenience of cities, &c. The first instance on record, that has been yet noticed, occurs where Herbert, Prince of Carrara, caused the first clock that was ever publicly exposed, to be erected at Padua. It was erected by John Dondi, whose family afterwards, in consequence, had the pronomen of Horologia assigned them, in remembrance of this circumstance: it is also mentioned on the tombstone of the artist. The family of Dondi now followed the profession of manufacturing clocks; for his son, John Dondi, constructed one upon improved principles.

The first clock at Bologna was put up in the year 1356. Some time after the year 1364, Charles V., surnamed the Wise, King of France, caused a clock to be placed in the tower of his palace, by Henry de Wyck, whom he had invited from Germany for the purpose, because there was then at Paris no artist of that kind, and to whom he assigned a salary of six sols per diem, with free lodgings in the Tower. Towards the end of that century, probably about the year 1370, Strasburg had a clock. About the same period, Courtray was celebrated for its clock, which the Duke of Burgundy carried away, A. D. 1382. A public clock was erected in the Altburg gate at Spire, in 1395, the works of which cost fifty-one florins.

The greater part of the principal cities of Europe, however, at this period, had clocks without striking. Clocks could not be procured but at a very great expense: of this, an instance occurred in the city of Auxerre, in the year 1483, when the magistrates being desirous of a clock, but discovering that it would cost more money than they thought themselves justified in expending on their own authority, applied to the Emperor Charles VIII. for leave to employ a portion of the public funds for that purpose.

In 1462, a public clock was put up in the church of the Virgin Mary at Nuremberg.

At Venice a public clock was put in the year 1497. In the same century an excellent clock was put up for Cosmo de Medici, by Lorenzo, a Florentine.

Having thus mentioned their origin in various places, until they came to ornament the religious houses, the palaces of kings, and the chief European cities, it now remains for us to take some notice of their existence in our own country for public use. From public documents still extant, it appears that so great was their expense considered in those early times of their introduction, that it was only the powerful and the rich who could procure them. We discover that the first clock for public and lay purposes in England was one erected on the north side of Old Palace Yard, Westminster, on which was this inscription, *Discite justitiam moniti*; which inscription is said to have been preserved many years after the clock-house had been decayed.

It is asserted that this clock was placed in that situation, for the purpose of being heard by the members of the courts of law; and the occasion which produced its existence is thus recorded. It was the produce of a fine levied upon the lord chief justice of the court of King's Bench, in the reign of Edward I. A. D. 1288, of whom it appears by a book called the "Year-Book," that this magistrate had been fined 800 marks for making an alteration in a record, wherein a defendant had been fined 13*s.* 4*d.*, and he, the chief justice, made it appear to be 6*s.* 8*d.* instead of that, the larger sum.

Notoriety, however, was attached to this transaction from the following circumstances. First, it appears to have been one of three questions put by Richard III. to his judges, with whom he was closeted in the Inner Star Chamber, to take their opinions on three points of law. The second question was, "Whether a justice of the peace, who had enrolled an indictment which had been negatived by the grand jury, among the true bills, might be punished for the abuse of his office?" On this question a diversity of opinion arose among the judges, some of whom supposed a magistrate could not be prosecuted for what he might have done; whilst others contended that he might, and cited the case of the lord chief justice above mentioned: so far was the answer of the judges strictly proper and historically true. The third circumstance to which we have alluded, and which is most material to our present question, is the application of the fine. It appears that it was expended in the construction of a clock, which was erected on the north side of Old Palace Yard; so that the judges, barristers, and students could not enter or leave the court, without having an opportunity of being reminded of the punishment of the chief justice, for presuming to violate the impartial duty of his high office; nor could they even hear it strike, whilst upon the throne of justice, without having his case repeated in their ears; thereby acting as a constant remembrancer, intimating they were to administer justice more than mercy.

Sir Edward Coke observes that 800 marks were actually entered upon the roll, so that it is extremely probable he had himself seen the record.

This clock was considered so important during the reign of Henry VI., that we find that the king gave the charge of keeping it, with its appurtenances, to William Warley, dean of St. Stephen's, with the pay of sixpence per diem, to be received at the exchequer.

The clock of St. Mary's, Oxford, was also furnished in 1523, out of fines imposed upon the students of that university.

With respect to the clock procured from the fine of the lord chief justice, we must also observe that its motto appears to relate to that circumstance; but though it might be said that it might relate to a dial as well as to a clock—a material observation to our present inquiry—yet, with respect to its present absence, it should be noticed, that it is probable that clock was a very indifferent one, but from its antiquity and the tradition attending it, was permitted to remain till the time of Elizabeth; then being quite decayed, a dial might have been substituted upon the same clock-house, bearing the very singular motto which, however originally applied, clearly alludes to such a circumstance as reported of the lord chief justice. This dial is placed on the very site where the clock-house stood.

But it is said by Derham, in his "Artificial Clockmaker," that the oldest clock in this kingdom is in Hampton Court Palace, marked with the letters N. O., presumed to have been the initials of the maker's name, of the date of 1540; but that author is evidently mistaken, in alleging that to be the oldest, because the Oxford clock bears a date seventeen years anterior to that period. With respect to the initials, or whatever they may be, we do not consider them of the smallest importance.

From Shakspeare's "Othello" it is proved that the ancient name of this instrument was Horologe; which various passages in our poets and old authors establish:—

"He'll watch the horologe a double set, If drink rock not his cradle."

Chaucer also says of a cock,

"Full sickerer was his crowing in his loge, As is a clock, or any abbey *orloge*;"

which tends to show that, in his time, clocks had been confined to religious houses.

So Lydgate's prologue to the story of Thebes:—

"I will myself be your orologere
To-morrow early."

With respect to our modern clocks, it would be presumption in us to say one word, as there is not an individual but knows as much about them, as we could tell him. We have fulfilled our intention in giving this historical account, which we are persuaded will afford some information. We will now proceed to

WATCHES,

WHICH are not of so great antiquity; as it is only about 1490, mention is made of watches, which first occurs in the Italian poems of Gaspar Visconti. Dominico Maria Manni says the inventor was Lorenzo a Vulparia, a native of Florence.

One might naturally be inclined to believe that the honour of original invention is duly demanded by the whole Germanic people, from the claim of the invention of watches being aspired to by the Nurembergians; as Doppelenayer gravely alleges they were first invented by a person residing in that city, in the sixteenth century, of the name of Peter Hale; and, perhaps, he has no better foundation for his conjecture, than that watches were at first of an oval shape, and were called Nuremberg eggs.

Shakspeare, in his "Twelfth Night," speaking of a *watch*, has the following expression, used by Malvolio: "I frown the while; and perchance wind up my watch, or play with some rich jewel." Also, the Priest, in answer to Olivia,

"Since when, my *watch* hath told me, toward my gravel have travelled but two hours."

The following observations appear to sanction our opinion of the early existence of those machines in this country. Dr. Derham, in his "Artificial Clockmaker," published in 1714, mentions a watch of Henry VIII., which at the period he wrote was in good order. Indeed, Dr. Demainbray says that he had heard Sir Isaac Newton and Demoire both speak of that watch.

An anecdote is related of the Emperor Charles V., contemporary with Henry VIII., which it appears has reference to the policy of Europe at that day. It is said, the emperor, after dinner, used to sit with several watches on the table, with his bottle in the centre. After the prince's retirement to the abbey of St. Just, he still continued to amuse himself with keeping them in order. From his inability to effect this correctly, it is reported he drew the rational reflection, *that it was impossible to effect what he had attempted—the regulation of the policy of Europe.*

It also appears that many watches of that day struck the hours. The "Memoirs of Literature" report that such watches having been stolen from Charles V. and Louis XI. whilst they were in a crowd, the thieves were detected from their striking.

It also appears from the evidence of certain watches of ancient construction formerly held by Sir Ashton Lever, and also by Mr. Ingham Forster, that *catgut* usually supplied the place of a chain in ancient watches; also that they were of a smaller size than now made, and generally of an oval form.

Imperfections of this nature, and probably other causes, might have rendered their truth uncertain, and this most probably precluded their general use, until the latter end of the reign of Elizabeth. The instances we have shown will prove they were generally known, and perhaps used at the time of Shakspeare writing the "Twelfth Night." And in the first edition of Harrington's "Orlando Furioso," published in 1591, the frontispiece represents the author with what appears to be a watch, although the engraving is extremely indistinct; moreover, the inscription to which engraving, of *Il Tempo passo*, clearly indicates the same thing.

Charles I., in 1631, incorporated the clockmakers company, and by charter, which prohibits clocks, watches, and alarums from being imported; which circumstance proves, that the English at this period, had no need of the aid of foreign ingenuity in this branch of mechanism.

We are told that Guy Fawkes and Percy were detected in the third year of James I., with a watch about them, which they had purchased, "to try conclusions for the long and short burning of the touchwood," (in the words of the time) which was prepared to give fire to the train of gunpowder.

The most material improvement introduced in this branch of mechanical knowledge took place in the addition of pendulums, by Huygens, as applied to clocks; for which conception he was indebted to Galileo, which that philosopher adopted for measuring time, he having taken the idea from observing the vibrations of a lamp in a church. This reign also boasts of the production of repeating-watches in England; first fabricated under the direction of the celebrated Dr. Hook, and manufactured by Tompion.

An anecdote is related of the attention paid to watches by James II., recorded by Derham, in the "Artificial Clockmaker:" One Barlow had procured a patent, in conjunction with the lord chief justice Allebone, for repeaters; but a person of the name of Quare making one at the same time, upon principles he had entertained before the patent was granted to Barlow, the king tried both in person, and gave the preference to Quare's, and caused it to be notified in the gazette.

In the next reign, the reputation of British watchmakers had increased so much, that an act was passed by parliament, enacting that British-made watches should be marked with the maker's name, in order to preserve the reputation of this branch of British manufacture from coming to discredit in foreign markets.

Thus we have given a general outline of the history of this branch of mechanics, for a period of nearly a thousand years, from the first invention of clocks by Pacificus of Verona, in 846, to the beginning of last century, since which period they have become an article of such general use to require no comment from us. We have noticed the various improvements in the order in which they occurred, among which the most striking feature appears to be the addition of the pendulums, as serving to regulate the motion of the machine; from its given length, certain weight and uniform vibration, it must be conceived to have been a happy thought in Galileo, for the admeasurement of time, and its application to this branch of mechanics was no less fortunate in Huygens. To discover the first invention of time, we will require to look back for upwards of two thousand years, at which period we will find

WATER-CLOCKS.

THESE are called Clepsydræ. Vitruvius, the Roman architect and mechanist, attributes the invention of the water-clock to Ctesibus of Alexandria, who flourished in the reign of Ptolemy Euergetes, about two hundred and forty-five years before the Christian era. The same author says, the machine was first introduced at Rome, two hundred and fifty seven years previous to the Christian era. There is reason to believe it was first introduced at Rome into courts of justice, from Greece, as it had been originally used in Greece for this purpose; the Roman orators being guided in the time they occupied the court, by this instrument, as we may learn from this expression of Cicero, "*Latrare ad clepsydram.*" Cicero also informs us, that it was first introduced into courts of justice, in the third consulate of Pompey.

It has been discovered that the inventions of Egypt, Chaldea, and other Oriental countries constantly travelled to Rome and the West. Long since the respective periods previously mentioned, has the honour of this invention been claimed by Burgundians, Bolognese, and other Italians; sometimes by Frenchmen, but chiefly by Germans.

Their claim for invention seems to be questionable in numerous instances, whatever it may be for improvement; they certainly cannot, consistently with what we have stated, be considered as the *first* inventors; although there is nothing to be alleged against these respective people being the discoverers of designs which had a previous existence unknown to them.

With equal or much more propriety might the Arabians, in point of time (could that be of consequence) be considered as inventors of this machine; and they are well known to possess the least claim to original invention of any people. They, however, have a merit, notwithstanding: but it is of a negative kind; for those arts, sciences, &c. which were (by chance) saved from the destruction of their bigoted ignorance, and which, when the fortune of war had thrown into their hands those pure designs of intellectual Greece, mere accident had wrested from their zealous fury. These they transmitted to a more ingenious people as pure as they had received them; but upon precisely as good grounds as the before-named Europeans claimed this *original* invention, might the Arabians have assumed that honour. For we read that Haroun al Raschid, Caliph of Bagdad, then the chief of the Saracen empire, sent as a present to Charlemagne, a clock of curious workmanship, which was put into motion by a clepsydra; which instrument is said, by Dr. Adams, "to have been used by the ancients to measure time by water running out of a vessel."

It consists of a cylinder divided into small cells, and suspended by a thread fixed to its axis in a frame, on which the hour distances, found by trial, are marked out. As the water flows from one cell into another, it changes slowly the centre of gravity of the cylinder, and puts it in motion.

The form of this instrument is thus described by Dr. Beckmann:—

"The most common kinds of these water-clocks, however, correspond in this, that the water issued drop by drop through a hole of the vessel, and fell into another, in which a light body, that floated, marked the height of the water as it rose, and by these means the time that had elapsed."

The most improved form the same instrument has acquired, is thus described, by the same author, from one in his own possession.

"Amongst the newest improvements added to this machine may be reckoned an alarum, which consists of a bell and small wheels, like that of a clock that strikes the hours, screwed to the top of the frame in which the cylinder is suspended. The axis of the cylinder, at the hour when one is desirous of being awakened, pushes down a small crank, which, by letting fall a weight, puts the alarum in motion. A dial plate with a handle is also placed over the frame."

In respect to the invention of clepsydræ, we should think the original inventor took his first idea from the use of an instrument common in Egypt, which that people called a *Canob*, or Nilometer, being a large stone vessel of the shape of a sarcophagus, into which water was daily poured, by proper officers, during the increase of the Nile, to show the people whether they had a prospect of plenty, or were to expect a scarcity in the ensuing year. As the fall of the water, after it had risen to a due height, was of equal importance to them; so the water was suffered to run out proportionably to its decrease in the river, being ascertained by just and equal marks which they generally well understood.

Vitrum horae had also been invented to describe the progress of time. These were conical hour-glasses, in which were placed a portion of sand; the glasses were joined together at the apex of the cone, with a small aperture of communication between the two.—From the glass, in which the sand is deposited, it dropped, grain by grain, into the sand below, standing upon its flat basis.

These machines are called hour-glasses, and well known. We have been unable to discover any account of the origin of this instrument; but, from its simplicity, it admits of no improvement. It is also believed this had its origin in a convent.

SPINNING.

THE necessity for human clothing must be so obvious, we should think, at nearly the first existence of our race, that two opinions upon that subject cannot exist. For, admitting the region where our first parents were stationed was more genial to life than these, our northern countries, yet the difference in temperature between the heat of noon-day, and the chilly damps of night, must be obvious to every one who has resided in, or has read of, tropical climates. Therefore, from necessity, we contend, our first parents could not have dispensed with the benefit of clothing. However, independent of the necessity of the thing, the Jewish History informs us that the first man, Adam, and his wife, in consequence of their unfortunate disobedience and positive violation of the commands of their Divine Creator, knew of their own nakedness; and, therefore, they were ashamed to answer to the sacred summons. This they confessed, with a simplicity congenial to truth, and in the same moment, frankly owned the cause; answering to the awful interrogatory of "Who told thee that thou wast naked? Hast thou eaten of the tree whereof I commanded thee that thou shouldest not eat?"—"The woman, whom thou gavest to be with me, she gave me of the tree, and I did eat." However, we are previously informed that, "the eyes of them both were opened, and they knew that they were naked; and they sewed fig-leaves together, and made themselves aprons."

It should be observed, that the leaf of the Banyan, or Indian fig, is probably here meant; if it is, the luxurious leaf of this tree is about three feet long, and proportionably wide; therefore, we may rationally conclude, much art was not required; probably a thorn might supply the place of a needle, and a blade of grass would do for a thread.

Afterwards, we are told, in the same chapter,—"Unto Adam, also to his wife, did the Lord Jehovah make coats of skins, and clothed them." The preceding is the earliest account of humanity; at the same time, it also furnishes the most ancient relation of the original of human clothing. From hieroglyphical inscriptions still extant, the most ancient inhabitants of Egypt wore sometimes clothing made of feathers, fastened together; sometimes of shells, also attached to each other; but the most general ancient clothing consisted of the skins of various animals. So is Hercules, and many of the heroes, clothed, in antique statuary. Although the sacred history is silent on this head, we may, perhaps, by inference, arrive at some clue or thread to guide us through the labyrinth of uncertainty.

Accordingly we find in the first passages, which will admit of constructive inference, that thread, of some sort, must, of necessity, have had existence:—"And Ada bare Jubal: he was the father of such as dwell in *tents*, and of such as have cattle."—Gen. iv., 20. Now, we submit, the inference of not only spinning, but also of weaving, and even sewing, must be conceded, before we can conceive the existence of tents. The cloth whereof they were made at that period, it is probable, was of the fleece of sheep; because of the early existence of woollen cloth among the Greeks, we have no doubt, from the following and numerous other passages in their poets; and also from the practice of Tyrian artisans, who were, we know, generally and confessedly eminent for their dying the imperial purple, and other scarce, valuable, and beautiful colours; and no substance better receives, or so well retains the most splendid of colours than does wool. But Homer speaks expressively in point, where, in his "Iliad," he expresses the truce which took place between the belligerent armies of Greeks and Trojans. After the defeat of Paris by Menelaus, and where the laughter-loving goddess, Venus, is said to have rescued her favourite from the fate he deserved to find; after she had conveyed the recreant hero from the field to his apartment, she then, like a true friend to matrimonial infidelity, goes in search of the Spartan queen, for the purpose of bringing the lovers together. She discovered the beautiful adultress on the walls of the city, where she had been describing to Priam, and his ancient nobles, the Trojan councillors, the various persons of the heroes of Greece. Upon this occasion, Venus, to use the language of the poet (as translated by Pope), assumes a disguise.

> "To her, beset by Trojan beauties, came,
> In borrowed form, the laughter-loving dame;
> She seemed an ancient maid, well skill'd to cull
> The snowy fleece, and wind the twisted wool."

The labours of Penelope, Helena herself, and innumerable passages in the works of the poet, all tend to confirm the fact.

That *linen* had also an early existence is proved at a very anterior period of the Jewish history. They had even fine linen previous to the construction of the utensils used in sacred worship; as, in Exodus, an ephod of linen is expressly mentioned; likewise in the xxvth chapter, 4th verse of that book, fine linen is expressly enumerated among those presents that the people were expected to offer freely to the Lord Jehovah. Whence we are justified in inferring they had most probably learned in Egypt to carry its structure to great perfection.

We have linen mentioned likewise, in Homer, upon the breach of the truce between the Grecians and Trojans with their auxiliary forces. On Menelaus having been wounded by an arrow from the bow of Pandarus, where the poet sweetly sings—

"But thee, Atrides, in that dangerous hour,
The gods forgot not, nor thy guardian power,
Pallas assists, (and weakened in its force),
Diverts the weapons from its destined course;
So, from her babe, when slumber seals his eye,
The watchful mother wafts the envenom'd fly.
Just where his belt, with golden buckles join'd,
Where LINEN folds the double corslet lin'd.
She turn'd the shaft, which, hissing from above,
Passed the broad belt, and through the corslet drove;
The folds it pierc'd, the plaited LINEN tore
And raz'd the skin, and drew the purple gore."

From what appears in the subsequent, as well as the former, part of this article, we submit, that the general manufacture of cloth, both woollen and linen, has been established; and if this is made out, the prior existence of the other subsidiary arts of spinning, weaving, &c. cannot be denied.

There are hieroglyphical symbols in the British Museum, which denote the various operations of the manufacture of cloths; and upon a monument upwards of three thousand six hundred years old.

Numerous arts have been discovered by mere accident. We are told, the very valuable operation of *feldt* making was discovered by a British sovereign, whose feet being always cold in the winter, he had wool put into his shoes; the moisture there contracted, the natural heat of the body, with the action to which this wool was exposed, between the foot and the shoe, caused the fleecy substance to consolidate; whence the origin of that very necessary article, the *Hat*.

STOCKING MANUFACTURE.

THE invention to which this article refers, affords a warm subject for panegyric. That clothing for the feet be warm, medical writers have in all ages recommended, and truly upon the most rational as well as philosophic and experimental practice; the feet, lying the most remote of any member from the heart, require, and particularly by people in years, to be kept warm, in order for their present comfort, as well as to promote the essential evacuation of superfluous humours, by perspiration, without which no frame can be healthy. So strongly is this precept impressed in our national moral habits, that it has formed a general maxim for the preservation of health. Even Thomas Parr is said to have observed, upon being asked to what cause he attributed the protraction of his life, "To keep the head cool by temperance, and the feet warm by exercise, to eat only when hunger required satisfaction, and to drink only when thirsty." We should suppose that this recipe would be at least worth a waggon load of the puffed quack pills which are palmed upon the public as made from a recipe left by that venerable man.

The art of knitting nets is one of great antiquity, as those nets used by the Hebrews, as well as by the Greeks, are conceived to be similar to those used in the present day. It was thought by Ovid, in his sixth "Metamorphosis," that the public were indebted to the spider for the origin of this ingenious invention; which would indeed seem probable, as it appears that the insect is prompted to be thus ingenious for the gratification of its natural wants, the web serving as a net or gin for the capture of flies and other small insects which supply it with food. And if our memory serve us, we recollect that the poet also, speaking of flies, observes that the web of the spider serves to secure the weak flies only, whilst the strong break it and escape; alluding to the influence of wealth and power to pierce through those laws which were made for the protection of the weak against the encroachments and violence offered by the strong. The author of Job, in the eighteenth chapter and ninth verse, mentions gins. However, in knitting stockings, the operation, as well as the effect, is essentially different from knitting nets. In the latter the twine is knotted into distinct meshes, which are secured by knots; in the former, the entire substance is produced without knots. To this distinction is to be ascribed the reason why knit stockings may become unravelled. In the other species the knots not only prevent the material being taken apart, but they also render the nets sufficiently strong to prevent even vigorous fish from escaping, yet being so capacious as to permit little fish to escape with the water.

The art of knitting is not now, by any means, so general as it was formerly. It then unquestionably rated among the number of female accomplishments; and it is certainly rather wonderful, because when the mechanism is once obtained, it requires no exertion of intellect to practise it; it may be carried on while sitting, walking, and talking, or in almost every situation to which ordinary life is called; and when it is considered that its produce adds to the comfort of the indigent, to the advantage of the poor,— and that to persons in easy circumstances habitual industry increases their happiness, these things considered, it is with wonder and regret we see it fallen into disuse; particularly as it is an occupation suited to every age and capacity, which the infant is strong enough to practise; and even in the infirmity and weakness of age it is practicable. We certainly do hope and trust these observations may invite the attention of those meritorious individuals who have the direction and management of our scholastic establishments, to revive the practice.

Fishing nets are also in use among the most barbarous nations of this period, as various navigators have satisfactorily proved; frequently made of rude materials, it is true—some of the bark of trees, and others of the beards of whales, besides a variety of other articles which the more refined inhabitants of civilised countries would never think of using for such a purpose.

The art of making nets, or ornaments of fine yarn, is said not to be a modern invention, it having been practised for hangings, and articles of dress and ornament. In the middle ages, it appears, the clergy wore netting of silk over their clerical robes. Professor Beckmann also says, he suspects those transparent dresses were used by ladies more than four hundred years ago, to cover those beauties they still wish to be visible.

The invention for making coverings for the legs, of this manufacture, is, we understand, of much later invention. It is well known that the Romans and the ancient nations had no particular covering for their legs. Indeed the necessity was not so urgent with the inhabitants of warm climates, as with those in our northern regions, who, we find, generally covered not only the feet, but the legs, thighs, and loins, with the same garment. Such, there is reason to conclude, were the trews, or trowsers, anciently worn by the Scotch, but not knit hose, which the following lines, from an old song, will help to prove:

"In days whan gude King Robert rang,
His trews they cost but half a croun:
He said they were a groat o'er dear,
And ca'd the tailor thief and loun."

A celebrated author on antiquities says, "It is probable the art of knitting stockings was first found out in the sixteenth century; but the time of the invention is doubtful." He continues, "Savary appears to have been the first person who hazarded a conjecture that this art is a Scottish invention, because when the French stocking-knitters became so numerous as to form a guild, they made choice of St. Fiacre, a native of Scotland, to be their patron; and besides this, there is a tradition, that the first knit stockings were brought to France from that country." This St. Fiacre, it appears, was the son of Eugenius, said to have been a Scottish king in the seventh century; and Fiacre lived as a hermit at Meaux, in France; in the Roman calendar, his name is opposite to the 30th of August.

More probable, however, is the opinion in this country which respectable writers support among them. We are informed by the author of the "History of the World," that Henry VIII., who reigned from 1509 to 1547, and who was fond of show and magnificence, at first wore woollen stockings; till by a singular occurrence he received a pair of silk knit stockings from Spain. His son Edward VI., who succeeded him on the throne, obtained by means of his merchant, Thomas Gresham, a pair of long Spanish knit silk stockings; this present was at that time highly prized. Queen Elizabeth, in the third year of her reign, A. D. 1561, received by her silk-woman, named Montague, a pair of knit silk stockings, and afterwards refused to wear any other kind.

Stowe also relates, in his "General Chronicle of England," that the Earl of Pembroke was the first nobleman who wore worsted knit stockings. In the year 1564, William Ridor, an apprentice of master Thomas Burdet, having accidentally seen, in the shop of an Italian merchant, a pair of knit worsted stockings, procured from

Mantua, having borrowed them, made a pair exactly like them; these were the first stockings that were knit in England, from woollen yarn. From this it would appear, that knit stockings were first introduced into England in the reign of Henry VIII., and that they were brought from Spain to this country; and probability appears to favour the belief that they were originally the produce of either that country or Italy. Should this be the case, it has been conceived by Professor Beckmann, that they came originally from Arabia to Spain.

The investigation with respect to the feigned productions of Rowley, published by the unfortunate Thomas Chatterton, arose from the mention of knitting, in a passage of those poems; it being contended that knit hose were unknown in the days of Rowley. The passage alluded to occurs in the tragedy of "Ella:"—

"She sayde, as herr whytte hands whytte hosen were knyttinge,
Whatte pleasure ytt ys to be married!"

A like ordeal took place with respect to Macpherson's Ossian from a similar reason, the mention of the sun's reflection setting on a glass window: now the existence of Ossian being contemporary with that of Julius Cæsar, it was contended that at that period it was not customary to glaze windows.

The Johnsonian faction set about that business in a very unsystematic manner: they should have procured some well qualified Erse scholar to have gone into those wilds where Macpherson declared he collected his materials from oral traditionary recitals, and have heard the poems themselves from the mouths of the aged inhabitants. If the traces of them could not have been found, they might then have ascribed the superior honour to Macpherson of writing a work that Greece, or Rome, in the splendour of literary glory, never surpassed, for many poetical beauties.

The people of Scotland, in the beginning of the sixteenth century, had, in the proper sense of the word, breeches; and wore a kind of stockings, their hose coming only to the knees; their stockings were made of linen or woollen, and breeches of hemp.

It is supposed that these particular articles of dress were also common in England, at and after that time, for in the year 1510, Henry VIII. appeared upon a public occasion, with his attendants, in dresses of the following description:—"The king and some of the gentlemen had the upper parts of their hosen, which was of blue and crimson, powdered with castels and sheafes of arrows of fine ducket gold, and the nether parts of scarlet, powdered with timbrels," &c. There may be occasion to suppose the upper parts of the hose were in separate pieces, as they were of different colours. Hollinshed, also speaking of another festival says, "The garments of six of them were of strange cuts, every cut *knit* with points of gold, and tassels of the same, their hosen cut in and tied likewise."

In A. D. 1530, the word *knit* appears to have been quite common in England, for John Palsgrave, a French master to the Princess Mary, daughter of Henry VIII., published a grammar, in which he stated, that this word in French was applied to the making of nets as well as of caps and of stockings.

In the household book of a noble family in the reign of Henry VIII., kept during the life of Sir Thomas L'Estrange, Knight of Hunstanton, Norfolk, by his Lady, Ann, daughter of Lord Vaux, there are the following entries, whence the price of those articles at that period are ascertained:—

1533. 25 H. 8. 7 Sept. Peyd for 4 peyr of knytt hose—viii *s.*

1538. 30 H. 8. 3 Oct. —— 2 peyr of knytt hose—i *s.*

It is observed that the first four pairs were for Sir Thomas, and the latter for his children.

Nevertheless, in the reign of Mary, i.e. 1558, many wore cloth hose, as is evidenced in the following anecdote of Dr. Sands, who was afterwards Archbishop of York. Being in the Tower, he had permission for a tailor to come and take an order for a pair of hose. This serves to prove the veracity of Stowe, that stockings were not an article manufactured in England generally, we suppose, till six years afterwards. "Dr. Sands, on his going to bed in Hurleston's house, he had a paire of hose newlie made, that were too long for him. For while he was in the Tower, a tailor was admitted to make him a pair of hose. One came into him whose name was Beniamin, dwelling in Birchin lane; he might not speak to him or come to him to take measure of him, but onelie to look upon his leg; he made the hose, and they were two inches too long. These hose he praied the good wife of the house to send to some tailor to cut his hose two inches shorter. The wife required the boy of the house to carrie them to the next tailor, which was Beniamin that made them. The boy required him to cut the hose. He said, 'j am not the maister's tailor.' Saith the boy, 'Because ye are our next neighbour, and my maister's tailor dwelleth far off, j come to you.' Beniamin took the hose and looked upon them, he took his handle work in hand, and said, 'These are not thy maister's hose, but Doctor Sands, them j made in the Tower.'"

In a catalogue of the revenues of the Bishop of St. Asaph, it is stated, "The bishop of that diocese was entitled, as a perquisite, upon the death of any beneficed clergyman, to his best coat, jerkin, doublet, and breeches. Item, his hose or nether stockings, shoes, and garters."

About 1557, knitting must have become common, for Harrison, in his description of the indigenous produce of this island, says, the bark of the alder tree was used by the peasants' wives for dying stockings which they had knitted.

Hollinshed also informs us, that about 1579, when Queen Elizabeth was at Norwich, "upon the stage there stood at one end eight small women children spinning worsted yarn, and at the other end as many knitting worsted yarn into hose."

Silk stockings are said, in consequence of their high price, for a long time to have been worn only upon grand occasions. Henry II. of France, wore them for the first time, on the marriage of his sister with the Duke of Savoy in the year 1559.

In the reign of Henry III. who ascended the throne in 1575, the consort of Geoffroy Camus de Pontcarre, who held a high office in the state, would not wear silk stockings given to her by a nurse who lived at court, because she considered them to be too gay. Anno 1569, when the privy-councillor Barthold von Mandelsoh, who had been envoy to many diets and courts, appeared on a week-day with silk stockings, which he had brought from Italy, the Margrave John of Austria said to him, "Barthold, I have silk stockings also; but I wear them only on Sundays and holidays."

The knitting stockings with wires, called *weaving*, has been thought to bear a resemblance to the wire work in screens of churches. However, the invention of the stocking loom is thought more worthy of attention, because it is alleged to have been the production of a single person, and perfected at one trial; his name, and the exact period is ascertained; and, because it is founded upon a similar incident to that of the beauteous Corinthian maid, elsewhere mentioned, as the introducer of painting in Greece; we bestow a particular attention upon this incident which produced the stocking loom, trusting our fair readers will favour us with their attention, when they are informed it is ascribed to Love.

It is a complicated piece of machinery, consisting of no fewer than two thousand pieces; it could not have been discovered accidentally, but must have been the result of deep combination and profound sagacity.

Under the usurpation of Cromwell, the stocking-knitters of London presented a petition, requesting permission to establish a guild. In this petition they gave a circumstantial account of their profession, of its rise, progress, and importance. No doubt can exist but that in this document the petitioners rendered the best, and probably a true account of the origin and progress of their trade, that of stocking weaving being then scarcely fifty years old. The circumstances they stated being then within memory, any misrepresentation would have militated against them, and could have been easily contradicted. In Deering's account of Nottingham, this petition is found. In that town the loom was first employed, where it has given wealth to many.

From this account it appears the inventor's name was William Lee, a native of Woodborough, a village about seven miles distant from Nottingham, in which the following passage occurs: "Which trade is properly styled frame work-knitting, because it is direct and absolute knit-work in the stitches thereof, nothing different therein from the common way of knitting, (not much more anciently for public use practised in this nation than this,) but only in the number of needles, at an instant working in this, more than in the other by a hundred for one, set in an engine or frame composed of above two thousand pieces of smith's, joiner's and turner's work, after so artificial and exact a manner that, by the judgment of all beholders, it far excels in the ingenuity, curiosity, and subtility of the invention and contexture, all other frames or instruments of manufacture in use in any known part of the world."

The inventor of this ingenious machine was heir to a considerable freehold estate, and a graduate of St. John's College, Cambridge. Being, it is said, deeply enamoured of a lovely young country-girl, who, during his frequent visits, paid more attention to her work, which was knitting, than to her lover or his proposals, he endeavoured to find out a machine which might facilitate and forward the operation of knitting, and by these means afford more leisure to the object of his affections to converse with him. Love, indeed, is confessed to be fertile in inventions, and has been the efficient passion which has perfected many inventions for which the gratitude of the world is due; but a machine so complex, so wonderful in its effects, would seem to require a longer time than was probably allowed, and a cooler judgment than a lover's to construct such mechanism. But even should the cause appear problematical, there cannot exist a doubt but the real inventor was Mr. William Lee, of Woodborough, in Nottinghamshire.

Deering says expressly, that Lee made the first stocking-loom in the year 1589; this account has also been adopted by various English writers. In the Stocking-weaver's Hall, London, is an old painting, in which Lee is represented pointing out his loom to a female knitter, who is standing near him; and below is seen an inscription with the date 1589, the period of the invention. "The ingenious William Lee, Master of Arts of St. John's College, Cambridge, devised this profitable art for stockings, (but being despised, went to France,) yet of iron to himself, but to us and others of gold; in memory of whom this is here painted."

Lee set up an establishment at Calverton, a village five miles from Nottingham, but met with no success. In this situation he showed his work to Queen Elizabeth; from that princess he requested some assistance, his work having embarrassed rather than assisted him; but instead of meeting with that remuneration to which his genius and invention so well entitled him, he was discouraged and discountenanced. It need not, therefore, excite surprise that Lee accepted the invitation of Henry IV. of France, who having heard of the invention, promised him a magnificent reward if he would carry it to France. He took nine journeymen, and several looms to Rouen, where he worked with much approbation; but the king being shortly after assassinated, and internal commotions taking place, the concern got into difficulties, and Lee died in poverty at Paris. A knowledge of the machine was brought back to England by some of the workmen who had emigrated with Lee, and who established themselves in Nottinghamshire, which still continues the principal seat of the manufacture.

During the first century after the invention of the stocking-loom, few improvements were made upon it, and two men were usually employed to work one frame. But in the course of last century the machine was very greatly improved. The late ingenious Mr. Jedediah Strut, of Belper, Derbyshire, was the first individual who succeeded in adapting it to the manufacture of *ribbed* stockings.

Estimating the population of Great Britain, say sixteen millions, and the average annual expenditure of each individual upon stockings and knit gloves at five shillings, the total value of the manufacture will be £4,000,000, and we consider this rather to be under than over the mark.

The effect of this invention was very late in making its appearance in Scotland. Till far on in the eighteenth century, the use of knitted stockings was universal. Mittens, or woollen gloves for the hands, and boot-hose, for drawing over the legs in riding, were also quite common, and all were wrought by the hand. The manufacture was carried on solely by women, the wives and daughters of farmers, generally, and the produce was sold as the means of bringing in a small revenue. The introduction of the stocking-loom to Hawick, in 1771, and the change of manners which took place about this period, soon put an end to this traffic; but still the greater part of the stockings worn by the country people on ordinary occasions are knitted at home. The art is also still in use in Shetland, where knitting forms the only amusement to relieve the tedium of a long winter, and where the articles produced are exceedingly fine in the texture: the Shetland hose bring the highest price of any woollen stocking.

COACHES.

COACH is said to be derived from *caroche*, Italian; a term first used in the eleventh century, and invented to designate a military machine, so called.

We intend the word coaches to stand for the generic name of all those machines used for the carriage of persons, on business or pleasure, (except, indeed, those for the conveyance of the dead,) from the state carriage of the sovereign down to the humble gig. The original inventor of this species of carriage is said to have been an Athenian monarch, 1489 years before Christ, who being afflicted with lameness in his feet, first invented a coach for his convenience, and with a view to conceal his debility. This may be regarded as the first original, of the kind, of Grecian invention.

The ancient historian, Diodorus Siculus, makes mention of a carriage in which Sesostris was wont to be drawn; and also, he says when he entered the city, or went out to the sacrifice, had four of his captive kings yoked to his chariot; but it is conjectured this carriage, to which that historian alludes, was a warrior's car. There is, most assuredly, ample room to believe that this was the first species of carriage which was introduced; if so, those existed long before the Athenian king above-named; because all the Homeric heroes, Greeks as well as Trojans, and their auxiliaries, rode in these machines, called chariots, or warriors' cars, which are also known to have existed long antecedent to that period. We remain assured that war chariots were used in the first ages of the world, by all the great monarchs who possessed dominion.

That species of carriage before said to have been invented by the Athenian monarch, we therefore presume, was a covered carriage, similar to that species designated in the twelve tables of the Roman law, and by them called *arcera*, which was said to be a carriage of the last presumed description, and mentioned as being intended for the conveyance of the infirm. To this species of carriage succeeded the soft *lectica*. But we will leave this part of our subject, and proceed towards our own times.

After the subversion of the Roman power, the northern sovereigns, who had become the barbarous and ignorant oppressors of our species, introduced and established, among other political regulations, the feudal system, as it was called, by which all property in land was held by certain fiefs, whereby the king, or, as termed, lord of the soil, let certain portions of the land to his nobles, military officers, and other great persons, generally often on condition of certain services required to be performed, called knights' service, and other military tenures; by which custom those tenants of the sovereign had to provide certain men and horses to serve him in his wars.— These first tenants, or vassals, afterwards underlet those lands to villains, so named, in contradistinction to the present recognised term, from their living in villages or hamlets, and other tenants, from whom, in their turn, similar services and certain provisions were required.— Thus the European world, which had become the prey of effeminacy and luxury, had, by this single important circumstance, their character so radically changed, that, like the mysterious power of the Cadmæan wand of Harlequin, wrought so uncommon a change in the morals of European society, that those who had formerly kept carriages, and wallowed in all the soft luxurious delicacy of Asiatic effeminacy, suddenly, or, at least, progressively, became a society of hardy equestrian veterans. Insomuch, that masters and servants, husbands and wives, clergy and laity, all rode upon horses, mules, or asses, which latter animals were chiefly used by women, monks, and other religious professors. The minister rode to court; the horse, without a conductor, returned to the stable, till a servant, regulated by the horologe, took him back to the court for his master. In this manner, we are assured, the magistrates of the imperial cities rode to council, till as late as the beginning of the sixteenth century; so that in the year 1502, steps to assist in mounting were erected by the Roman gate at Frankfort. The members of the council who, at the diet and other occasions, were employed as ambassadors, were, on this account, called *rittmeister* in the language of the country; at present the expression riding-servant is preserved in some of the imperial cities. The entry of great lords in public into any place, or their departure from it, was never in a carriage, but always on horseback; in all the pontifical records, speaking of ceremonials, no mention is made either of a state coach, or body coachman, but of state horses and state mules. In the following regulation, it is found that the horse which his Holiness rode "was necessary to be of an iron-grey colour; not mettlesome, but a quiet, tractable nag. That a stool of three steps should be provided for the assistance of his Holiness in mounting: that the emperor, or kings, if present, were obliged to hold his stirrup, and lead the horse."

Bishops made their public entry, on induction, on horses or asses richly caparisoned. At the coronation of the emperor, the electors and principal officers of the empire were ordered to make their entry on horseback.—It was formerly requisite, that those who received a fief, or other investiture, should make their appearance on horseback. The vassal was obliged to ride with two attendants to the court of his lord, where, after he had dismounted his horse, he received his fief.

Covered carriages were again introduced in the beginning of the sixteenth century, for the accommodation of women of the very first rank; the men, however, thought it disgraceful to ride in them. At that period, when the electors, and other Germanic princes, did not choose to be present at the meeting of the States, they excused themselves to the emperor, that their health would not permit them to ride on horseback, which was considered as an *established point*, that it was unbecoming to them to ride like women. What, according to their prevailing ideas, was not permitted to princes, was much less allowed to their servants. In A. D. 1554, when Count Wolf, of Barby, was summoned by John Frederic, Elector of Saxony, to go to Spires, to attend the convention of the States assembled there, he *requested leave*, on account of ill health, to make use of a close carriage with four horses When the counts and nobility were invited to attend the solemnity of the elector's half brother, John Ernest, the invitation was accompanied with a memorandum, that such dresses of ceremony as they might be desirous of taking with them, should be transported in a small waggon;—which notice would have been unnecessary, had coaches been generally used among those nobles. The use of covered carriages was in fact, for a long time, prohibited even to women, the consorts of princes. About the year 1545, the wife of a certain duke obtained from him, with great difficulty, permission to use a covered carriage in a journey to the baths, in which permission there was this express stipulation, that none of her attendants were to be permitted this indulgence: though much pomp was displayed upon the occasion by the duchess. Such is the influence of example in our superiors, who can mould dependents and inferiors to whatever shape they please.

Notwithstanding all these ceremonious regulations, about the end of the fifteenth century, kings and princes began to employ covered carriages in journeys, and afterwards on public solemnities. When Richard II., towards the close of the fourteenth century, was compelled to fly from his rebellious subjects, himself with all his followers, were on horseback; but his mother, who was weak and sick, rode in a carriage. But this became afterwards unfashionable here, for that monarch's queen, Anna, daughter of the King of Bohemia, showed the English ladies how gracefully she could ride on a side-saddle; and therefore whirlicotes (the ancient name for coaches in England), and chariots, were disused in England, except on coronations and other public solemnities.

In the year 1471, after the battle of Tewkesbury, which decided the fate of Henry VI., and that of the house of Lancaster, when others flew in different directions, the queen was found in her coach, almost dead with sorrow.

In 1474, the Emperor Frederic III. came to Frankfort in a close carriage; and as he remained in it on account of the wetness of the weather, the inhabitants had no occasion to support the canopy which was to have been held over him, while he went to the council house and returned. In the following year, the same emperor visited that city in a very magnificent carriage. In 1487, on occasion of the celebration of the feast of St. George at Windsor, the third year of Henry VII., the queen and king went in a rich chaise; they were attended by twenty-one ladies. In the description of the splendid tournament held by the Elector of Brandenburg, at Ruppin, in 1509, Beckmann says, he reads of a carriage all gilt, which belonged to the Electress;

of twelve other coaches, ornamented with crimson; and of another, belonging to the Duchess of Mecklenburgh, which was hung with red satin.

In the Northumberland household book, about this period, is an order of the duke for the chapel stuff to be sent before in my lord's chariot.

At the coronation of the Emperor Maximilian, 1562, the Elector of Cologne had twelve carriages. In 1594, when John Sigismund did homage at Warsaw, for Prussia, he had in his train thirty-six coaches, with six horses each. Count Kevenhiller, speaking of the marriage of Ferdinand II. with a princess of Bavaria, says, "The bride rode with her sisters in a splendid carriage studded with gold; her maids of honour in carriages hung with black satin, and the rest of the ladies in neat leather carriages."

Mary, Infanta of Spain, spouse of Ferdinand III., rode, in 1631, in a glass carriage, in which no more than two persons could sit. The wedding carriage of the first wife of the Emperor Leopold, who was a Spanish princess, cost, with the harness, 38,000 florins. The coaches used by that emperor are thus described:—"In the imperial coaches no great magnificence was to be seen; they were covered over with red cloth and black nails. The harness was black, and in the whole work there was no gold. The panels were of glass, and on that account they were called the imperial glass coaches. On festivals the harness was ornamented with red silk fringes. The imperial coaches were distinguished only by their having leather traces; but the ladies in the imperial suite were obliged to be content to be conveyed in carriages, the traces of which were made of ropes." At the magnificent court of Ernest Augustus, at Hanover, there were in 1681, fifty gilt coaches, with six horses each. So early did Hanover begin to surpass other cities in the number and splendour of its carriages.

The first time that coaches were introduced into Sweden was towards the end of the sixteenth century, when John of Finland, among other articles of luxury, brought one with him on his return from England.

Beckmann also informs us, that the great lords of Germany first imagined that they could suppress the use of coaches by prohibitions. There is still preserved an edict, in which the feudal nobility and vassals are forbidden the use of coaches, under pain of incurring the punishment of felony.

Philip II., Duke of Pomeranian-Stettin, reminded his vassals also, in 1608, that they ought not to make so much use of carriages as of horses. All these orders and admonitions, however, were of no avail, and coaches became common all over Germany.

Persons of the first rank (ladies we presume), in France, frequently sat behind their equerry, and the horse was often led by servants. When Charles VI., wished to see, *incognito*, the entry of the queen, he placed himself behind his master of the horse, with whom, however, he was incommoded in the crowd. Private persons in France, physicians, for instance, used no carriages in the fifteenth century. In Paris, at all the palaces and public places, there were steps for mounting on horseback.

Carriages, notwithstanding, appear to have been used very early in France, as appears by an ordinance issued in 1294, for suppressing luxury, and in which the citizens were prohibited from using carriages. About 1550, there were at Paris, for the first time, only three coaches; one of which belonged to the queen; another to Diana of Poictiers, the favourite mistress of two kings, Francis I. and Henry II.; and the third to René de Laval, a corpulent nobleman, unable to ride on horseback. Henry IV. was assassinated in a coach; but he usually rode through the streets of Paris on horseback. For himself and his queen he had only one coach, as appears by a letter which he writes to a friend, which is still preserved: "I cannot *wait* upon you to-day, because my wife is using my carriage."

Roubo, in his costly treatise on joiners' work, has furnished three figures of carriages used in the time of Henry IV., from drawings preserved in the King's Library: from them it is seen those coaches were not suspended by straps, that they had a canopy supported by ornamental pillars, and that the whole body was surrounded by curtains of stuff or leather, which could be drawn up. The coach in which Louis IV. made his public entrance about the middle of the seventeenth century, appears from a drawing in the same library to have been a suspended carriage.

Our national chronicler, John Stowe, says coaches were first known in England about 1580; he likewise says, they were first brought from Germany by the Earl of Arundel, in 1589. Anderson places the period when coaches began to be used in common here about 1605. It is remarked of the Duke of Buckingham, that he was the first who was drawn by six horses, in 1619. To ridicule this pomp, the Earl of Northumberland put eight horses to his carriage.

Things are altered now when we have carriages of every description—for the high and low, the rich and the poor. Vis-a-vis,—an open carriage chiefly constructed for the benefit of conversation, as its name implies. Landau, landaulets, phætons, chaises, whiskeys, cabs, fiacres, &c., &c., are but names adapted to different purposes, and constructed nearly upon the same principles as coaches, but some of them close, others open, some to be opened or shut according to the weather, or taste of the passengers, and calculated to contain an indefinite number, from two to six persons; nay, there are the jolly good omnibuses running in every town and village in the kingdom, the generality of which are constructed to carry twelve inside and eight outside passengers.

The number of hackney coaches which ply in the streets of London have been augmented from time to time, since their first establishment in 1625, when there were only twenty. Coaches, cabs, omnibuses, &c., now plying, amount to nearly three thousand.

To prevent imposition, the proprietors of these carriages are compelled to have their names painted on some conspicuous place of the carriage, and their number affixed in the inside, as well as the out. This regulation has become absolutely necessary of late years, on account of the numerous frauds practised by the coachmen.

We read that in Russia there are employed clumsy, but very convenient sorts of carriages, so constructed as to be either closed or open, and to hold a bed or couch, called *brichka*, with which persons can travel even for two or three thousand miles without much inconvenience, except it be over the rough stones of their towns, owing to the superior accommodations of either lying down or sitting; this change of position renders a journey less irksome, without which it would prove intolerable. In Russia, from Riga to the Crimea, at least, post horses are furnished by the government, and entrusted to subalterns in the Russian army to provide them.

Coaches for hire were first established by public authority in France, as early as 1671. There are employed in the streets of the capital no fewer than three thousand hackney coaches. As early as the year 1650 Charles Villerme paid into the royal treasury fifteen thousand livres, for the exclusive privilege of keeping and using fiacres in Paris.

Post chaises were introduced in the year 1664.

Hackney coaches were established in Edinburgh in 1673, when the number was only twenty. Public fiacres were introduced at Warsaw in 1778. In Amsterdam the coaches have no wheels; nor have they any at Petersburg in the winter—they are used as sledges.

The state-coach of the city of London is a species of heir-loom, or the hereditary property of the city; it is a very large and apparently extremely heavy machine, but superbly decorated with large panels of crystal glass, richly gilt, and elegantly painted with several appropriate designs. In one of the centre panels, among a group of figures, is one supporting a shield bearing the inscription "*Henry Fitzalwin*, 1189," in the old English, character; therefore we conjecture that the coach was constructed at a period coeval with the above date.

SADDLES, BRIDLES, AND STIRRUPS.

IN the earliest ages it was customary to ride without either bridles or saddles, if the poet be worthy of credit; for we observe Lucan, speaking of the Massillians, says:

"Without a saddle the Massilians ride,
And with a bending switch their horses guide."

They regulated the motion of the horses by a switch and their voice. It has been observed, that the case was the same with the Numidians, Getulians, Libyans, as well as most of the Grecian people. As the reason of the thing appears to point out the superior expediency of a bridle, they afterwards came into fashion among the Greeks, which they called *lupi*; because it is said the bit of the bridle bore a resemblance to the teeth of the wolf, whence Lucan says of it:—

"Nor with the sharper bitsManage th' unruly horse."

In the east it would appear that bridles, at least, were used at an early period. For we have a great number of texts in the Scripture, which definitely express as much: in the Psalms, and likewise in Proverbs, the name and application of the bridle is often particularly mentioned, and more frequently alluded to. Virgil, indeed, says, referring to very early times:

"The Lapithæ of Pelethronium rodeWith bridles first,—and what their use was show'd."

The saddle is also of ancient origin, for we read in I. Kings, xiii., 13.,—"And he said unto his sons, *saddle* me the ass. So they saddled him the ass: and he rode thereon." And before that period, in the second generation after Noah, the Assyrian empire was established. In its commencement, even as early as the days of Semiramis, the wife of Ninus, the first Assyrian king, who built Ninevah, there were those articles of horse furniture, called *packs* and *fardles*; for in ancient historians we find the following passage occur in this respect. "Semiramis ascended from the plain to the top of the mountain, by laying the packs and fardels of the beasts that followed her, one upon another." The same author informs us that this was Mount Bagistan, in Medea, and that it was seventeen furlongs from the top to the bottom.

In the first ages, among the Greeks and Romans, a cloth or mattrass, a piece of leather or raw hide, was all they used for a saddle. Such coverings afterwards became more costly: Silius Italicus says, they were made of costly skins.

It, however, appears, that after they were become common, it was considered as effeminate to use them; hence the Romans despised them: and in his old age, Varro boasts of having, when young, rode without a covering to his horse. Xenophon reproaches the Persians, because they put more clothes upon the backs of their horses than upon their beds. From the aspect in which hardy people viewed this practice, the warlike Teutones considered it most disgraceful, and despised the Roman cavalry.

In the fifth century, saddles were so magnificent, that a prohibition was issued by Leo I., that they should not be ornamented with pearls or precious stones. In the sixth century, the Emperor Mauritas directed that they should have coverings of fur, of large dimensions.

From every information we have been able to collect, we believe that the appendage of stirrups were not added to saddles before the sixth century. It is said, that previous to the introduction of stirrups, the young and agile used to mount their horses by vaulting upon them, which many did in an expert and graceful manner; of course, practice was essential to this perfection. That this should be afforded, wooden horses were placed in the Campus Martius, where this exercise was performed of mounting or dismounting on either side; first, without, and next with arms. Cavalry had also, occasionally, a strap of leather, or a metallic projection affixed to their spears, in or upon which the foot being placed, the ascent became more practicable. Respecting the period of this invention, Montfaucon has presumed that the invention must have been subsequent to the use of saddles; however, opposed to this opinion, an ingenious argument has been offered, that is possible they might have been anterior to that invention; because, it is said, they might have been appended to a girth round the body of the horse. Both Hippocrates and Galen speak of a disease to which the feet and ancles were subject, from long riding, occasioned by suspension of the feet without a resting-place. Suetonius, the Roman, informs us that Germanicus, the father of Caligula, was wont to ride after dinner, to strengthen his ancles, by the action of riding affording the blood freer circulation in the part.

The Latin names assigned them have been various, among which is *scalæ*; in which sense Mauritius, in his treatise on the art of war, is said to have named them. Now, this writer is supposed to have lived in the sixth century; but we conceive it is pretty evident they had an earlier existence in Arabia, Turkey, and Persia, as there is an alto, as well as bas-relief of this last country, still extant, which is believed to have been as ancient as the days of Darius, because it was brought from the city he built, Persepolis, having this representation.

The invention and name of stirrup is supposed to have been borrowed from the anatomy of the ear, where a band is found resembling it in form.

HORSE-SHOES.

WHEN we consider the vast importance of security to the feet of that useful animal, the horse, we cannot but feel surprised that on account of the very rough roads the ancients must occasionally had to travel, that some metallic shoes had not been invented and introduced previously to the period when they appeared.

That the security of the rider necessarily depended upon the safety of the animal he rode, cannot be questioned. Hence, then, we do not wonder to observe, that the sagacious Aristotle and Pliny should remark upon the covering placed upon the feet of those animals of draught and burden. From what these authors have said, however, we dare not conclude that the feet of horses or camels were faced or shod with iron: but it should rather seem that in time of war, or on long journeys, the feet of both kinds of beasts were prepared with such species of shoes as the common people wore, and which were generally made of strong ox-leather. We are told that when the hoofs of cattle, particularly oxen, had sustained any injury or hurt, they were furnished with shoes made of Spanish or African broom, with which linen is often manufactured in the south of France and Italy; also shoes of some of the plants of the hemp kind, which were woven or plaited together. Although these may be considered as only a species of surgical bandages with regard to oxen; but such shoes were particularly given to mules, which in days of old were employed much more than at present for riding; and from some instances of immoderate extravagance in people of rank, it appears that they had for their animals very costly shoes of some of the most valuable metals. Nero, when he undertook short journeys, was drawn always by mules shod with silver, and those of his wife were shod with gold.

The circumstance being barely mentioned, without any particular detail, we are anxious to afford any certain information on the mode in which those shoes were constructed. From a passage in Dio Cassius, we have reason to believe that it was only the upper part of the shoe that was made of those costly metals, or that they were plaited from thin slips.

Xenophon relates that a certain people in Asia were in the habit of drawing socks over the feet of their horses, when the snow lay deep on the ground. The Kamschatkian employs the same means to preserve the feet of his dogs, which draw his sledge, or hunt the seals upon the ice. Those species of shoes, according to Captain Cook, are so ingeniously made as to be bound, and at the same time to admit the claws of the animal through them.

From a passage found in Suetonius, we may infer that the Roman horse-shoes were put on in the manner we have mentioned; for that author says, that the coachman of Vespasian once stopped to put on the shoes of his mules: this being the case, the probability appears pretty certain, that in deep roads and moist soils the animals must have frequently lost their shoes.

Artemedorus speaks of a shod horse, and uses the same kind of expression whilst speaking of other cattle. Winkelman has described a cut stone in the collection of Baron Stosch, on which is represented the figure of a man holding one foot of a horse, whilst another, kneeling, is employed in fastening a shoe.

That it was not usual to shoe the war-horse, may be gathered from this,—when Mithridates was besieging Cyzicus, he was obliged to send his cavalry to Bythnia, because the hoofs of the horses were entirely spoiled and worn out. Diodorus Siculus informs us, that Alexander, in his expedition, proceeded with uninterrupted marches, until the feet of his horses were entirely broken and destroyed. A like instance occurs in Cinnamus, where the cavalry were obliged to be left behind, because the horses had suffered considerably in their hoofs, to which he adds, they were often liable. Hence it may, perhaps, appear, that such horse-shoes as are now in use, were unknown to the ancients; and Chardiu gives no representation of them in ancient Persian antiquities. In the grave of Childeric, a northern chieftain and King of France, was discovered a piece of iron, which the learned antiquarians who saw it, pronounced, from that portion of it which the rust had left, to have been an old horse-shoe; they saw, or thought they saw, four distinct apertures for nails on each side; but whilst they were endeavouring to remove the corrosive excrescence of rust, to ascertain with more certainty, it broke under their hands. The reason why we mentioned this here is, that if the relic discovered was really a horse-shoe, it must have been one of the most ancient specimens known; because, we find that monarch died in the year 481; his grave was discovered at Tournay in 1683. The occasion of his having a horse-shoe in his grave, was from the creed of his religion; the superstitious belief of the Scandinavians taught them to place implicit confidence in the power of this amulet, to prevent the ingress of evil spirits. The remains of this belief is even now often seen in the obscure streets of the British metropolis; and, indeed, throughout the country, where the mystic shoe frequently appears as the faithful guardian of the domestic threshold.

It is, we understand, the opinion of the French historian, Daniel, that, in the ninth century, horses were not shod always, but only in the time of frost, and on some other very particular occasions.

The practice of shoeing horses was introduced into England by William I. We are told that this monarch gave the city of Northampton as a fief to a certain person, one of his attendants, in consideration of his paying a certain sum yearly for the shoeing of horses. And it is also alleged, that Henry, or Hugh de Ferres, or de Ferrers, was the same person who held this fief on the above condition, and who was the ancestor of the family of that name, and who still bear six horse-shoes in their coat of arms. This was the person whom William entrusted to inspect his farriers.

We should not omit to observe, that it is remarked, that horse-shoes have been found, with other riding furniture, in the graves of some of the old inhabitants of Germany, and also in those of the Vandals in the North of Europe.

GUNPOWDER.

THE express period when *nitrum* was first discovered is extremely uncertain; but that this nitrum is an alkaline salt, there is little difficulty in proving. It has, indeed, been conjectured that it was a component part of the *Greek fire*, invented about the year 678, which has been generally believed to be the origin of gunpowder. From the oldest prescriptions which have been found, and which is said to be that given by the Princess Anna Commena, in which, however, only resin, sulphur, and oil are mentioned, saltpetre does not appear.

It is believed by an author very well qualified to form a judgment on the question, that the first certain account we have of saltpetre by that express name, occurs in the oldest account of the invention of gunpowder, which, according to him (Professor Beckmann) occurred in the thirteenth century. Dr. Rees, in his Cyclopedia, expressly says, about the year 1320; and that it was first used by the Venetians employed against the Genoese in 1380; also that it was first in Europe at a place now known as Chrogia, against Laurence de Medicis; and the last named authority adds, "That all Italy made complaint against it, as a contravention of the law of arms." Dr. Rees gives the following recipe for its manufacture, without distinguishing the proportionate parts:—"A composition of nitre, sulphur, and charcoal, mixed together, and usually granulated." He describes its effects by observing, that "it easily takes fire, and when fired, rarifies and expands with great vehemence by means of its elastic force;" also that "it may be made without *nitre*, by means of *marine acid*."

We have two accounts preserved to us of the original of this invention. The first of which was given by our illustrious countryman, Roger Bacon, called the Wonderful Doctor, who died A. D.1278; previous to which period, gunpowder must have existed. The other account is by Albertus Magnus, in a work published in 1612.

It is said to be doubted whether Albertus was the author of the book which bears his name; but that he, whoever he may have been, and Bacon, are presumed to have taken their information from the same identical source. About the period of the invention of gunpowder, it appears the art of making the Greek fire began to be lost. In the works of Roger Bacon, the term occurs three times. According to Casiri, the term *pulvis nitratus*, is to be found in an Arabic MS. the author of which existed about 1249. If the work of Geber, *De Investigatione Perfectionis*, be genuine, and if this writer lived, as has been thought, in the eighth century, it would be the oldest where saltpetre is mentioned, in a prescription for an *aqua solutiva*, which appears to be almost *aqua regia*.

We are inclined to believe, however, from various authorities, that gunpowder was invented in India, as it was proved in a paper read before the French National Institute, by M. Langles, that the Arabians obtained a knowledge of gunpowder from the Indians, who had been acquainted with it from the earliest periods. The use of it in war is said to have been prohibited them in their sacred books. It was employed in 690 at a battle near Mecca, by the Arabians. It was brought by the Saracens from Africa to the Europeans, who improved the preparation, and first discovered various ways of employing it in war. In no country could saltpetre and its various uses be more easily discovered than in India, where the soil is so rich in nitrous particles that nothing is necessary but lixiviation to obtain saltpetre; and where this substance is so abundant, that almost all the gunpowder used in different wars, with which European sovereigns have tormented themselves, burdened their subjects with intolerable taxes, and cursed the world from its invention—has been made from Indian saltpetre. Had not saltpetre been known previous to the thirteenth century, neither could gunpowder or aquafortis have existed; and for the best of all reasons, that neither of them could be made without saltpetre or nitre. But should it appear that this neutral salt was known in India long prior to that period, and used by Indians as well as Arabians before they were employed by Europeans, and considering the former to have practised chemistry previous to the latter; should this have been proved, perhaps a similar proof will necessarily await upon the articles aquafortis and gunpowder. Because if this affirmation be established, it will be discovered that Europeans knew nothing of aquafortis until after the Arabian chemists.

Probability appears to favour the idea, that at or about the twelfth century the accumulated number of consequents, from the improvement in European science, the arts we now possess were introduced into our catalogue, i. e., nitre, aquafortis, and gunpowder.

After the period that saltpetre became necessary to governments for the manufacture of gunpowder, they endeavoured to obtain it at a cheap rate; and for that purpose were guilty in some countries of the most violent and oppressive measures, intruding upon private property of every description to furnish it, hunting for the effervescence even in old walls, to the great annoyance of individuals. But after repeated acts of the most flagrant oppression from the public officers, and from farmers, to whom this iniquitous practice was entrusted, they could not procure a sufficiency; but were obliged to have recourse to traffic in India for that purpose.

GUNS.

THAT these dangerous weapons were not known in Europe previous to the introduction of gunpowder may be safely inferred; as without that substance their necessity or utility is wanting.

At first the construction of this machine was characterised by that awkward, rude, and cumbersome appearance which generally distinguished all inventions in their infancy; reminding us of those very rude instruments brought from the Sandwich Islands, and deposited in our Museum.

The first portable fire-arms were discharged by a match; in course of time this was fastened to a cock, for the greater security of the hand whilst discharging the piece. Afterwards a fire-stone was attached, screwed into a cock, with a steel plate before it, and fixed in a small wheel, which could be wound up by a key, affixed to the barrel. This fire-stone was not at first of a vitreous nature, like that now in use for striking fire, but a compact pyrites, long known as such, and called a fire-stone. As an instrument so furnished was often liable to miss fire, till a late period a match was still continued with the wheel; and it was not till a considerable time after that, instead of a friable pyrites, so much exposed to effloresce, a vitreous stone was affixed to the improvement of the lock, somewhat resembling our own gun-lock. But these progressive improvements advanced slowly, because as recently as the early part of the last century these clumsy contrivances were in use. During that period, those instruments were denominated by various names, chiefly German and Dutch, such as *buchse*, *hakenbuchse*, *arquebuss*, musket, martinet, pistol, &c. The first of these names arose from the oldest portable kind of fire-arms having a similarity to a box. There were long and short *buchse*, the latter of which were peculiar to cavalry; the longest kind also, from their resemblance to a pipe, were called in Germany, *rohr*.

Large pieces, which were conveyed on carriages, were called *Karren buchse*, from the action of conveyance. Soon afterwards cannon were introduced, at first called *canna*; now known as artillery. However, artillery-men, and others concerned in those employments, still use the terms previously mentioned. The hackenbuchse were so very large and unwieldy, that if carried in the hand, they could not be used manually alone; they were, therefore, supported by a post or stay, called a *bock*, because it had a forked end, somewhat resembling the horns of the buck, between which the piece was fixed by a hook projecting from the stock. There is still preserved in the Tower of London, an old *buchse*; a specimen of every species of our national arms may be seen in the same place.

From those terms before-mentioned, it would appear, that not only the English, but also the French, and most other European nations, took the names of their fire-arms.

It appears that pistols were first used in Germany; they had a wheel attached to them. Bellay mentions them in the year 1544, in the time of Francis I.; and under Henry II., the German horsemen were called *pistoliers*. Several historians think that the name came from Pistolia, in Tuscany, because there they were first made; and, if we might hazard an opinion, we think this conjecture right. Hence, although Germany might first have generally used them, we think they were an Italian invention.

Muskets are said to have received their name from either the French *mouchet*, or else from the Latin *muschetus*; however, we are of opinion that neither of these terms gave its original; and submit that it is derived from the Latin *muscarium*,—the fall of men being as sudden after the explosion of this deadly weapon, as the death of a fly after it is flapped by that instrument, which was common in the butcher's shambles of ancient Rome.

Daniel proves they were known in France as early as the period of Francis I. Brandome, however, asserts they were introduced by the Duke of Alva—that cruel monster in human shape—that tool of a blood-thirsty tyrant—whose name has its full merit when it has eternal execration, as the exploits of that diabolical character in the Spanish Netherlands bear indubitable testimony: that wretch existed in 1507; and they were not known in France at that period, as Brandome endeavours to prove, or we should have had more intelligence handed down to posterity by the commentators of one who would so willingly have used such an instrument. The *lock* is said to have been invented in the city of Nuremberg, in Germany, about 1517; but that cannot be considered as the lock of the present day, as even in Germany the fire-lock is known by the name of the French-lock, which certainly militates against the previous assertion, the one giving the name perhaps to the other.

Beckmann says, "In the history of the Brunswick military it is stated, that the soldiers of that Duchy first obtained flint-locks instead of match-locks in 1687. It has often been asserted," he continues, "that fire-tubes which took fire of themselves were forbidden first in Bohemia and Moravia, and afterwards in the whole German empire, under a severe penalty, by the Emperor Maximilian I.; but I have not found any allusion to this circumstance in the different police laws of that emperor."

That the first fire-stones were pyrites appears from various sources, and afterwards a vitreous kind of stone was introduced in its stead; this circumstance is said to have produced some kind of confusion, as in many instances the properties were applied to that stone which were related by the Germans of antiquity as belonging to pyrites. In Germany, this vitreous stone was called *vlint*; in Sweden and Denmark, *flinta*; and in England, *flint*. This appellation is of great antiquity.

Anciently, in Germany, as it appears from the song of Hildebrand, a metrical romance of very early date, that Hildebrand and Hudebrand, a father and son, and, at the moment, ignorant of their affinity, agreed to fight for each other's armour; and it is said "They let fly their ashen spears with such force, that they stuck in the shields, and they thrust resounding axes of flint against each other, having uplifted their shields previously; but the Lady Ulta rushed in between them—'I know the cross of gold,' said she, 'which I gave him for his shield; this is my Hildebrand. You, Hudebrand, sheath your sword; this is your father!' Then she led both champions into the hall, and gave them meat and wine with many embraces."

Besides these proofs that the ancient name of the stone was known in Germany by the appellation *vlint*—which species of stone may, perhaps, without hazarding the danger of error, be conceived to be the same which Zipporah, the wife of Moses, is represented to have used, in the 25th verse of the 4th chapter of Exodus: "Then Zipporah took a sharp stone, and cut off the foreskin of her son, and cast it at his feet, and said,—Surely a bloody husband thou art to me." And it is added she said so, on account of the circumcision.

In addition to what has appeared, let us add, it cannot be doubted that the instrument fired by this stone first obtained for it, in Germany, the name of *vlint*; as the ancient name may, in general, be now lost, it is commonly called flint-stone. Those people acquainted with the northern, Scandinavian, and German antiquities, know that the knives employed in ancient sacrifices, and other sharp instruments, were made of this stone, as appears from the remains being yet discovered in old barrows, and between urns.

It is also presumed that the Ethiopian stone, mentioned as used by one of the Egyptian embalmers, first to open the body to get at the intestines, was a flint-stone. The soil being in some places siliceous or chalky, naturally produces such stones in common with that earth.

The flint is a stone indigenous in most European countries; they are commonly collected and manufactured by people whose occupation allows them much spare time. The easiest mode to shape them is with a species of pillow of saw-dust, or some other soft material, sown up in coarse cloth, held upon the knees, and with a hammer having a bevil edge, they may be broken into almost any form or size by those accustomed to the practice.

The great quantity of the material from whence they are composed allows for any waste which accident may produce. In several counties of England they are so plentiful, that they are the common material employed for mending the public roads. But we are informed that this is not the case in France, where, in time of war, the people were prohibited from exporting them. The Dutch are commonly large dealers in this article.

Flint is a large component in the manufacture of glass.

Gun flints are now, however, comparatively little used, as percussion caps are generally substituted, which act with more certainty, and require a great deal less trouble.

ASTRONOMY.

NOT being greedy of delusion ourselves, neither would we lead others into error; but, on the contrary, are desirous to avoid all deception, as we may be considered over studious to give the most rational origin, and where we cannot get at the history of those objects which engage our attention—whenever this is uncertain we resort to nature, experience, and reason, and furnish the most correct explanation our contracted circle of information will permit. Whenever we discover the clue of history, we collect the most satisfactory detail our limits will afford us to insert. Guided by the preceding notions, and directed by those principles, we have endeavoured correctly to conceive, and faithfully to portray our own conceptions in the best manner our experience might enable us, to make a just distinction between metaphorical allusion and literal application; ever endeavouring to discriminate between serious assertion and studied fable.

We fully coincide with the just remark of the learned author of "Indian Antiquities," who says "that in respect to the early ages of the world, all the remains of genuine history, except that contained in the sacred annals, is only to be obtained through the mazes of Mythology."

It must be confessed, that to sift this grain of corn from the bushel of chaff with which it is surrounded, where every effort which the ingenuity of Greece could devise to render fable as current as truth, was resorted to, is no small task; that it requires the operation of the best exercised reason, and the assistance of extraordinary judgment, which is only to be attained through the medium of extensive experience and the exercise of clear and discriminative powers: then we pretend not to possess the best of possible acquisitions of this kind, but the best in our power, we have endeavoured to collect, and summoned to our assistance; and the value of our labours we will leave the public to judge.

If the application of observations like the preceding ever come *apropos*, surely they apply to the present article; since from the *sideral* science, all the errors of an idolatrous race proceeded in the major part of the population of the ancient world: from thence also proceeded the most sublime imagery which embellishes the syren voice of poetic song, the grandest metaphors, and the sweetest allegories, which ornament the transendent eloquence of the most able rhetoricians of Greece and Rome; the fire of exquisitely natural and most noble allusions which enliven and embellish their historic pages. The sweetest philosophical explications also flowed from thence, which ornament the various immortal works of their most excellent poets, orators, historians, natural and moral philosophers; and, in brief, of every description of the sublimest genius of ancient Greece and Rome, in their most divine effusions.

It will appear, we believe, that the first astronomers of Chaldea, Phœnicia, and Egypt, are not now known as astronomers, by name, if we except the person of the royal Nimrod, the founder of the Chaldean empire, which name is often confounded with Belus; sometimes one is put for the other, and often Belus is called the son of Nimrod. How the truth of this was, we shall not at present determine: be it as it may, it is allowed on all hands that the sideral science claims for its inventor no less a person than the founder of the first monarchy in the world. That this science was first introduced by the founder of the Tower of Babel is not questioned, because it is more evident, that in that country there must have existed from necessity, the expediency of the most approved observation, which could be made upon this eminently useful science; where, on account of the excessive solar heat, people generally travel by night: where, for hundreds of miles, are nothing but pathless deserts, with a horizon as boundless and little impeded as that of the ocean; assuredly under such circumstances, the local situation of the site of the immense Observatory of Babel must point out the expediency of procuring some intelligence from the position which the inhabitants discovered the host of heaven to appear in at the rising, setting, &c.; for from what will appear in the course of this article, it will be very evident that the Tower of Babel was constructed for the purpose of an astronomical observatory; farther, that the climate of Chaldea was most favourable to the exercise of that sublime art, will not admit of a question, when we consider the atmosphere is so pure, so clear, so free from exhalation, that at night the sky is said to resemble an immense canopy of black velvet studded with embossed gold, from the appearance of the stars; and that it was not only the appearance of the stars, their rising, setting, and motion, by which they knew time was to be measured; but also the distinction between one star and another could be correctly ascertained from the usual colour—here it was the various planets, zodiacal constellations, and the other asterisms in both hemispheres, received their primary names.

The preceding circumstance, it is conceived, fixes the local place where the science had its origin.

The Tower of Babel was a parallelogram, with sides of unequal length. Herodotus thus describes it.—"The Temple of Jupiter Belus occupies the other [square of the city], whose huge gates of brass may be seen. It is a square building; in the midst rises a tower of the height of one furlong, upon which resting as a base, seven other turrets are built in regular succession. The ascent is on the outside, which, winding from the ground, is continued to the highest tower: in the middle of the whole structure there is a convenient resting place."

Diodorus Siculus says, this tower was decayed in his time; but, in his description of Babylon, he thus speaks of it—describing it as the act of Semiramis, who flourished two thousand nine hundred and forty-four years before Christ:—"In the middle of the city, she built a temple to Jupiter-Belus; of which, since writers differ amongst themselves, and the work is now wholly decayed through length of time, there is nothing that can with certainty be related concerning it; yet it is apparent it was of an exceeding great height; and that, by the advantage of it, the Chaldean astrologers exactly observed the rising and setting of the stars. The whole was built of brick, cemented with bitumen, with great art and cost. Upon the top she placed three statues of beaten gold, of Jupiter, Juno, and Rhea: that of Jupiter stood upright, in the posture as if he was walking; it was forty feet in height, and weighed one thousand Babylonish talents. The statue of Rhea was of the same weight, sitting on a golden throne, having two lions standing on either side, one at her knees, and near to them were two exceeding great serpents of silver, weighing thirty talents each. Here, too, the image of Juno stood upright, and weighed eight hundred talents, grasping a serpent by the head in her right hand, and holding a sceptre adorned with precious stones in her left. For all these deities there was placed a table made of beaten gold, forty feet long and fifteen broad, weighing five hundred talents, upon which stood two cups, weighing thirty talents, and near to them as many censers, weighing three hundred talents: there were likewise placed three drinking bowls of gold—the one to Jupiter weighed two hundred talents, and the others six hundred each."

We have been thus circumstantial in our description of Babylon, for obvious reasons. First—that it was the first local situation where, since the deluge, men had associated for civil purposes; and secondly—because it was the original station where the astronomical science was cultivated. From Chaldea, Astronomy travelled to Egypt, where she was studied for many ages; she also went to Phœnicia, where she was regarded with equal attention. But the peculiar occasion which the Phœnician people had to improve their acquaintance with this science, will appear, upon reflecting that these people occupied a narrow and barren tract of land between the Mediterranean and Arabian seas; therefore, they found it essentially necessary to improve their situation by those means which Divine Providence had apparently marked out for them to resort unto; we accordingly find them applying to mercantile industry; as a commercial people, in this character, they were the ready medium of communication between every part of the then known world. In consequence, they had factories or mercantile stations up the Mediterranean; but particularly on its European side, on the shores of the Atlantic, and even in the British sea: we recognise their occupying Marseilles, and others, on the coast of France; Cadiz, on that of Spain; the Lizard Point, and other places, in Cornwall, where they traded for tin in the British Isles. In brief, their commercial spirit carried them to every part of the globe: by the by, admitting that rational belief be allowed to Plato and Solon, we shall find that they had, in the first ages, explored the Atlantic Ocean, and even discovered America. A great variety of authorities may be adduced to prove the assertion—that the Phœnicians made three descents on the American coast; and others, who say that the inhabitants discovered there by the Spaniards, gave the same names to the plants as had been assigned them in Asia; that their religious rites were similar, and general customs and manners the same,—we refer to Joseph Da Costa's "History of the Indies," published in 1694.

This author was an eye-witness, and wrote from actual observation. The Phœnicians, in the exercise of their mercantile functions, had the most obvious necessity to cultivate the sideral science. We find that they accordingly did so, and made various improvements and very important discoveries by their exercise. From the northern hemisphere being more known to them than it was to the Chaldeans, they discovered that splendid and beautiful asterism, *Cynosuræ*, or the polar-star,—an asterism of the most singular service, before the properties of the magnet were discovered, and which star was sometimes called, from them, Phœnice.

From Phœnicia and Egypt the celestial science of astronomy was brought into Greece, with which people the Phœnicians were intimate; for they, by trade, having occasion to converse with the Greeks, and also from uniting in one national resemblance, the three opposite characteristics of soldiers, sailors, and men of science, the communications between the two people were very frequent. At every period, from the first establishment of the Grecian states, that highly eminent and intellectual people collected from all others every particular they could obtain in all matters having relation to sciences and arts; those they cultivated with a success worthy of the motive which first induced them to make these collections.— Loving Knowledge for herself, they succeeded beyond all others in obtaining her favours.

The first Greek who appears on record to have cultivated the celestial science with success, was Thales, born at Miletus, in Asia Minor, six hundred and forty one years before Christ; he explained the causes of eclipses, and predicted one. He also taught that the earth was round, and divided into five zones; he discovered the solstices and equinoxes, and likewise divided the year into three hundred and sixty-five days. He had travelled into Egypt in search of knowledge, where he ascertained the height of one of the pyramids, from its shade. He looked upon water as the principle of all things. From him the sect called the Ionic had their origin.

Anaximander, his pupil, followed him, and supported the opinions of his great master; he was born before Christ six hundred and ten years; he invented maps and dials, and is said to have constructed a sphere. His ideas of the planets were, however, erroneous.

Anaximenes was a scholar of Anaximander, and born five hundred and fifty-four years before Christ. He taught that air was the origin of all things, and many erroneous notions; among others, that the earth was a plane, and the heavens a solid concave sphere, with the stars affixed to it like nails.

Anaxagoras of Clazomene, the pupil of, and successor to, Anaximenes, born before Christ five hundred and sixty years. The doctrines he supported are a strange association of important truths, mixed with the most gross absurdities. He taught that the world was made by a being of infinite power; that mind was the origin of motion; that the upper regions, which he called ether, were filled with fire, that the rapid revolution of this ether had raised large masses of stone from the earth, which, being inflamed, formed the stars, which were kept in their places, and prevented from falling by the velocity of their motion.

His ideas of the solar orb were extremely erroneous; alleging, according to different authors, various uncertain positions respecting the materials of which that planet is composed: one says, *he* said it was a vast mass of fire; another states *his* opinion, that it was red-hot iron; and a third, that it was of stone. He taught that the comets are an assemblage of planets; that winds are produced in consequence of highly rarified air; that thunder and lightning are a collision of clouds; earthquakes, by subterraneous air forcing its passage upwards; that the moon is inhabited, &c.

This philosopher removed his school from Miletus to Athens, which was thenceforth the grand seat of all learning. He had taught there for thirty years, when he was prosecuted for his philosophical opinions, particularly for his just ideas relative to the Deity, and condemned to death. When sentence was pronounced, he said:—"It is long since Nature condemned me to that." However, according to the laws of Athens, he was permitted an appeal to the people, in which his scholar, the immortal Pericles, saved his life by his eloquence. His sentence of death was changed into banishment. Whilst in prison he determined exactly the proportion of the circumference of the circle to its diameter, denominated "squaring the circle." He died at Lampsacus. Archelaus, his scholar, was the preceptor of the divine Socrates.

Pythagoras was another scholar of Thales. The place of his nativity is uncertain; but having settled in the island of Samos, he is generally reckoned of that place. He travelled in search of knowledge through Phœnicia, Chaldea, Egypt, and India; however, meeting with little encouragement on his return to Samos, he passed over to Italy, in the time of Tarquin the Proud, and opened a school at Croto, a city in the Gulf of Tarentum, where he had a number of students, and gained much reputation. His pupils were obliged to listen in silence for at least two years; if talkative, longer;

sometimes, for five years, before they were permitted to ask him any questions; for which time they were *mathematicoi*, because they were set to study geometry, dialling, music, and other high sciences, called by the Greeks *mathemata*. But the name of *mathematici* was commonly applied to those who cultivated the stellary science, and who predicted the fortunes of men, by observing the stars under which they were born.

This luminary of science first assumed the appellation of *philosopher*; before him, those whose pursuits have now that title, were called sages or wise men; he was the founder of the sect called the Italic. He was so much honoured whilst living, and his memory honoured when dead, by the Romans, that they attributed to him the learning of Numa, who lived much earlier. About the year of the city 411, the Delphian oracle having directed the Romans to erect statues to the bravest and wisest of the Greeks, they conferred that honour upon Alcibiades and Pythagoras.

He taught publicly that the earth is the centre of the universe; but to his scholars he gave his real opinions; similar to those afterwards adopted by Copernicus, that the earth and all the planets moved round the sun, as their co-centre, and which doctrine he is presumed to have derived from either the Chaldeans or Indians. He thought that the earth is round, and everywhere inhabited. Hence, he admitted that we might have antipodes, which name is said to have been invented by Plato.

Pythagoras was distinguished for his skill in music, which he first reduced to certain firm principles, and likewise for his discoveries in geometry. He first proved, that in a right-angled triangle, the square of the hypothenuse, or side subtending the right angle, is equal to the two other sides; also that of all plain figures having equal circumference, the circle is largest; and of all solids having equal surfaces, the sphere is the largest. Pythagoras likewise taught that all things were made of fire. That the Deity animated the universe, as the soul does the body; which doctrine, with that of the metempsychosis, or transmigration, he likewise taught; and which thoughts were adopted by Plato, and are most beautifully expressed by Virgil; that the sun, the moon, the planets, and fixed stars, are all actuated by some divinity, and move each in a transparent solid sphere in the order following:—next to the Earth, the Moon, then Mercury, Venus, the Sun, Mars, Jupiter, Saturn; the sphere of the fixed stars last of all; that those move with a sound inconceivably beautiful, which ears cannot comprehend. Those eight spheres he imagined to be analogous to the eight notes in music.

Empedocles, the chief scholar of Pythagoras, entertained the same sentiments with his teacher, concerning astronomy. He is said to have thrown himself into the crater of Mount Etna, to make himself pass for a god; or, perhaps, which may approach nearer the truth, because he could not discover the cause of the eruption: or else in his endeavours to discover the cause. One of his iron sandals being thrown up by the volcano, revealed the mode in which he had perished.

Philolaus, also a scholar of Pythagoras, first taught publicly the diurnal motion of the earth upon its axis, and its annual motion round the sun; which first suggested to Copernicus the idea of that system which he established.

Meteon, born at Leuconæ, a village near Athens, first introduced into Europe the Lunar Cycle, consisting of nineteen solar years, or nineteen lunar years, and seven intercalary months. It had been first adopted by the Chaldeans. Meteon published it at the Olympic games, where it was received with so great applause that it was then universally adopted through the Grecian States, and their colonies, and got the name of the Cycle, or Golden Number, to denote its excellence, which name it still retains.

It was also called the Great Year; which name was likewise applied to various spaces of time by different authors; by Virgil, to the solar year, to distinguish it from the monthly revolution of the moon; by Cicero and others, to the revolution of six hundred years, or three thousand six hundred years; called also several ages, when all the stars shall come to the same position, with respect to one another, as they were in at a certain time before; called likewise *Annus Mundanus*, or *Vertens*.

The lunar cycle begun four hundred and thirty-two years before the commencement of our era, and according to it, the Greek calendars, which determined the celebration of their annual feasts, &c. were adjusted. Meteon is said to have derived his knowledge of this subject from Chaldea.

The opinions of the subsequently registered astronomer, Xonophanes, founder of the Eleatic school, are so truly monstrous, that after the light which had appeared, he must have travelled with his eyes shut; or else the rage for novelty alike affected the scientific of Greece, as it did their *literati*; choosing to travel a long way for new thoughts, when they might have found much better at hand. Xonophanes, among other whimsical opinions, maintained that the stars were extinguished every morning, and illuminated every evening; that the sun is an inflamed cloud; that eclipses happen by the extinction of the sun, which is afterwards lighted up; that the moon is ten times larger than the earth; that there are many suns and moons to illumine different climates.

The Eleatic school was chiefly famous for the study of logic, or the art of ratiocination, first invented by Zeno. Those of this sect paid but little attention to science, or the study of Nature. Philosophy was anciently divided into three parts, natural, moral, and the art of reasoning. Xonophanes was succeeded by Parmenides, his scholar, who, in addition to his master's absurdities, taught that the earth was habitable in only the two temperate zones; that the earth was suspended in the middle of the universe, in a fluid lighter than air; that all bodies left to themselves light on its surface. This bore a slight resemblance to the Newtonian doctrine of attraction.

Democritus, of Abdera, a scholar of Leucippus, who flourished four hundred and fifty-six years before Christ, was the first publisher of the Atomic Cosmogony, invented by Mochus, the Phœnician, said to have been received by his master Leucippus. Both admitted plurality of worlds. Democritus was the first who taught that the milky way is occasioned by the confused light of an infinite number of stars; which doctrine is still maintained by the best informed of philosophers. He also extended that idea to comets; the number of which Seneca says the Greek philosophers did not know; and that Democritus suspected there were more planets than we could see. This was also the opinion of many others, the truth of which has been verified in the discoveries of Pallas, Juno, Vesta, and the *Georgium Sidus*.

Democritus is considered as the parent of experimental philosophy; the greatest part of his time was devoted to it; and he is said to have made many discoveries. He, like Meteon, and Newton, maintained the absurd idea of the existence of a vacuum, which was denied by Thales and Descartes. Democritus also maintained that the sea was constantly diminishing. He declared that he would prefer the discovery of one of the causes of the works of Nature, to the possession of the Persian monarchy. Often laughing at the follies of mankind, he was thought by the vulgar to be out of his mind; but Hippocrates, being sent to cure him, soon found him to be the wisest man of the age; and Seneca reckons him the most acute and ingenious of the ancients, on account of his many useful inventions; particularly his ingenious making of artificial emeralds, tinging them of any colour; of softening ivory, dissolving stones, &c.

Although the chief attention of Plato and Aristotle was directed to other grand objects, yet they much contributed to the improvement of astronomy. Notwithstanding the most famous in this respect was Eudoxus, the scholar of Plato, who was famous for his skill in astrology, natural and judicial, or the art of foretelling future events by the relative situations of the stars, of their various influences, an art which prevailed for many ages among the ancients, and is yet assiduously cultivated by the modern Arabians and other orientals, although in a great measure exploded in European nations. By the former or which divisions in this science are foretold the changes of seasons, rain, wind, thunder, cold, heat, famine, diseases, &c., from a knowledge of the causes that are believed to act upon the earth and its atmosphere; whilst the latter foretold the characters, fortunes, &c., of men, from the stellary disposition at the moment of their respective nativities.

The philosopher, Eudoxus, spent much of his time on the top of a high mountain, to observe the motion of the stars. He regulated the Greek year as Cæsar did the Roman. Had the ancient Grecian astronomers been equally attached to experiment with Democritus, they might have arrived at more certain conclusions; but they were content with speculative theory, and spoke rather from conjecture than observation; whence both Strabio and Polybius treated as fabulous the since recognised assertion of Pythius, a famous navigator to the north, who had sailed to a country supposed to be Iceland, where he said the sun, in the middle of summer, never set.

The most important improvements in astronomy were made in the school of Alexandria, founded by Ptolemy Philadelphus; and which seminary flourished for nine hundred and twenty-three years, till the invasion of the Saracen army, under the command of Amrou. Those astronomers were chiefly Greeks, or of Grecian extraction—the most learned men being invited here by the liberality of the Ptolemies. The first who distinguished themselves were Timocarus and Aristillus, prior to the foundation of the library, which was founded three hundred years before Christ. Those two men endeavoured to determine the places of the different stars, and thus to trace the course of the planets. The next and most eminent man was Aristarchus, about two hundred and sixty-four years before Christ; who taught, that the sun was about nineteen times further from the earth than the moon (which is not the twentieth part of its real distance), although the philosophers of the Pythagorean school did not consider it above three times, and perhaps only one and a half further distant. Aristarchus also taught, that the moon was fifty-six diameters of our earth from this globe, which opinion comes near to the truth: he believed it to be scarcely one-third of its real size. He was widely erroneous in his conception of the sun's dimensions. He also, in conformity to the doctrines of Pythagorus and Philolaus, supposed the sun to be placed in the centre, and that the earth moved round it; on which account he was accused of impiety, as disturbing the repose of the Vesta and the Lares. This opinion was not, however, retained by his successors in the Alexandrian school. Contrary to the doctrine of the Greek philosophers, he taught that the stars were at different distances, and that the orbit of the earth round the sun was an insensible point, in consequence of the immense distance of the stars. The only work of Aristarchus which remains, is on the magnitude and distance of the sun and moon.

Very nearly contemporary with Aristarchus was Euclid, the celebrated geometrician of Alexandria; Manetho, an astrologer and historian; and Aratus and Cleanthus, disciples of Zeno, the stoic philosopher; all of whom contributed to the enlargement of astronomical knowledge; but particularly the two first named.

Eratosthenes, born at Cyrene, succeeded Aristarchus, being invited by Ptolemy Euergetes. This professor is said to be the inventor of the Armillary sphere, an instrument or machine composed of moveable sides, representing the equator, the two colures, with the meridian; all of which turned round on an axis directed to the two poles of the world, each of which circles were anciently called armilla, and the whole machine, astrolabus. All instruments which could be contrived for the promotion of this science, were furnished at the public expense, and placed within the observatory of Alexandria. Assisted by these instruments, Eratosthenes first undertook to measure the obliquity of the ecliptics, or rather the double of that obliquity, that is, the distance from the tropics, which he made to be about 47 degrees; the obliquity, or half of this distance, 23½ degrees. This grand attempt was to ascertain the exact distance of a degree of the meridian, and thus to determine the circumference of the earth; which he accomplished with wonderful exactness, considering the period at which he lived; and he performed this by the same method since adopted by the moderns who have succeeded him. He is also said to have discovered the true distance of the sun from the earth.

The great Archimedes lived contemporary with Eratosthenes, that eminent geometrician of Syracuse, whose inventive genius in mechanics had constructed engines which protracted the fall of that capital, with its Island Sicily, to the almost omnipotent power of Rome for a considerable period.

The most illustrious astronomer which had as yet appeared at Alexandria was Hipparchus, who flourished between one hundred and sixty and one hundred and twenty-five years before Christ. He first brought this science into a tangible elementary form, rendering it systematic. He discovered, or was the first who observed the difference between the autumnal and the vernal equinox; the former being seven days longer than the latter, which proceeds from the eccentricity of the earth's orbit, first discovered from observing the inequality of the solar motion. He framed tables for what is called equation of time, or to ascertain the difference between the shade on a well constructed dial and a perfectly regulated clock. He made great progress in explaining the motions and phases of the moon; however, he was not so successful with respect to the planets.

His greatest work was his ascertaining the number of the stars, marking their distances, and arriving at the means by which their precise places on the hemisphere of Alexandria could be known. He marked one thousand six hundred stars, in seventy-two signs, into which the heavens were divided. Pliny says this was a labour which must have been difficult even to a god. The appearance of a new star induced him to set about and accomplish this work, which he did in a catalogue for the benefit of future observers.

Hipparchus does not mention comets, whence it has been conjectured he had never seen any; it has also been suggested, that he considered them with meteors, which are not objects of astronomical observation. He divided the heavens into forty-nine constellations, viz., twelve in the ecliptic, twenty-one in the north, and sixteen in the south. To one of these he gave the name of Berenice's Hair, in honour of the wife of Ptolemy Soter, who had consecrated her hair, which was very beautiful, to Venus Urania, if her husband should return from a war in Asia victorious; it being hung up in the temple of the goddess, soon after disappeared, and is said to have been carried off by the gods.

Hipparchus likewise constructed a sphere, or celestial globe, on which all the stars visible at Alexandria were depicted; and thought to have been similar to the Faranese globe at Rome, still extant. In his observations on the stars, he discovered that, when viewed from the same spot, their distance always appeared the same from each other; but he discovered the distance of the moon to be different in various parts of the heavens; for instance, in the horizon and zenith. This he conceived to be owing to the extent of the globe; he, therefore, contrived a method of reducing appearances of this kind, to what they would be if viewed from the centre of the earth, which is called a parallax; and the discovery of it was of the greatest importance to astronomy. He took this idea from observing that a tree, in the middle of a plain, appeared in different parts of the horizon, when viewed from different situations; so does a star appear in the various points of the heavens, when viewed in different parts of the globe. Hipparchus was the first who connected geography with astronomy, and this fixed both the sciences on certain principles.

After the overthrow of the Roman empire, the first encourager of learning was Charles the Great, or Charlemagne; but little could be done in his time; after his death the former ignorance prevailed. Beda, or Bede, from his piety and modesty termed *venerabilis*, and his scholar, Alcinius, both Englishmen, greatly excelled in general literature; among other qualifications they were eminent in the astronomy of the preceding period. The first step towards the revival of knowledge, or the translation of the Astronomical Elements of Alfergan, the Arab, by order of Frederick II., chosen Emperor of Germany in 1212. About the same time Alphonso X., King of Castile, assembled from all parts the most famous astronomers, who at his desire, composed what are called the Alphonsine Tables, founded on the hypothesis of Ptolemy.

About the same period John Sacrobosco, of Holywood, a native of Halifax, in Yorkshire, who was educated at Oxford, and taught mathematics and philosophy at Paris, made an abridgment of the amalgamist of Ptolemy, and of the commentaries of the Arabs, which was long famous as an elementary book under the title of "De Sphira Mundi." He died at Paris, in the year 1235. In the same year, Roger Bacon, an English Franciscan friar, made astonishing discoveries in science for the time he lived. He perceived the error in the Kalendar of Julius Cæsar, and proposed a plan, for the correction of it, to Pope Clement IV. in 1267. He is presumed from his writings to have known the use of optical glasses, and the composition and effects of gunpowder. He believed in planetary influence on men's fortunes, and the transmutation of metals. On account of his vast knowledge in astronomy, mathematics, and chemistry, he was called Doctor *Mirabilis*; but, for the same reason, he was suspected of magic. Under this pretext, whilst at Paris, he was put in prison by order of the Pope's legate; and after a long and severe confinement, he was at last, by the interest of several noble persons, liberated, returned to England, and died at Oxford in 1292, in the seventy-eighth year of his age.

In the fifteenth century two events happened which changed the face of the sciences; the invention of printing, about 1440, and the taking of Constantinople by the Turks in 1453. The learned men of that city having escaped from the cruelty of the victors, fled into Italy, and again introduced the taste for classical literature; which was greatly promoted by the munificence of the Emperor Frederick III., Pope Nicholas V., and particularly of Cosmo de Medici, who justly merited the title of Father of his Country, and Patron of the Muses.

The restoration of astronomy began in Germany. The first who distinguished himself, was George Purbach, born at Purbach, on the confines of Austria and Bavaria, in 1423, who was cut off in the flower of his age—only thirty-eight years old. He was succeeded by a scholar more skilful than himself, John Muller, born at Konigsberg, in 1436, who taught mathematics and astronomy with great reputation at Vienna. In February, 1471, appeared a comet, on which he published his observations. Being called to Rome by Pope Sextus IV., to assist in correcting the Kalender, he was cut off by the plague, in 1476. Bernard Waltherus, a rich citizen of Nuremberg, his friend and associate, succeeded him, who is said to have first made use of clocks in his astronomical observations, in 1484, and to have been the first of the moderns who perceived the effects of the refraction of light.

Fracastorius, born at Verona, in 1483, was a celebrated astronomer, and an eminent poet and good philosopher; he made considerable discoveries in this science, and with all his abilities may be considered as the precursor of the celebrated Copernicus.

Nicholas Copernicus, the restorer of the Pythagorean philosophy, and the modern discoverer of the rational and true system of astronomy, as now universally received, under the title of his name, was born at Thorn, a city of Royal Prussia, 19th February, 1473. Having learnt the Latin and Greek Languages in his father's house, he was sent to Cracow, to be instructed in philosophy and physic, where he was honoured with the degree of doctor; showing a greater predilection for mathematics than medicine. His uncle by his mother's side was a bishop, who gave him a canonry upon his return from Italy, whither he had gone to study astronomy, under Dominic Maria, at Bologna, and had afterwards taught mathematics with success at Rome. In the repose and solitude of an ecclesiastical life, he bent his chief attention to the study of astronomy. Dissatisfied with the system of Ptolemy, which had prevailed fourteen centuries, he laboured to form a juster one. What led him to discover the mistakes of Ptolemy was his observations on the motions of Venus; he is said to have derived his first notion on this subject from various passages in the classics, which mention the opinions of Pythagoras and his followers, as, indeed, he himself acknowledges in his address to Pope Paul III. He established the rotation of the earth round its axis, and its motion round the sun; but to explain certain irregularities in the motion of the planets, he retained the epicicles and eccentrics of Ptolemy. His work was first printed at Nuremberg, in 1543, a short time before his death.

The doctrines of Copernicus were not at first generally adopted. The most eminent professors in Europe adhered to the old opinions.

Among the astronomers of this period, the Landgrave of Hesse deserves particular praise, who erected a magnificent observatory at the top of the Castle of Cassel, and made many observations himself, in conjunction with Christopher Rothman and Justus Burge, concerning the place of the sun, of the planets, and of the stars.

But the person who enriched astronomy with the greatest number of facts of any modern who had yet appeared, was Tycho Brahe, a Dane of noble extraction, born in 1546, designed by his parents for the study of the law; but attracted by an eclipse of the sun in 1560, at Copenhagen, whither he had been sent to learn philosophy, he was struck with astonishment in observing that the phenomenon happened at the very moment it had been predicted.

He admired the art of predicting eclipses, and wished to acquire it. At first, for want of proper instruments, he fell into several mistakes, which, however, he afterwards corrected. Having early perceived his future improvements must depend on instruments, he caused some to be constructed larger than usual, and thus rendered more exact. On the 11th November, 1572, he perceived a new star in Cassiopeia, which continued without changing its place till spring 1574, equal in splendour to Jupiter or Venus. It last it changed colours and entirely disappeared. Nothing similar to this had been observed since the days of Hipparchus.

Tycho, in imitation of that illustrious astronomer, conceived a design of forming a catalogue of the stars. To promote his views, the King of Denmark ordered a castle to be built in Hueun, an island between Seonia and Zealand, which Tycho called Uranibourg, "the city of heaven," and where he placed the finest collection of instruments that had ever yet appeared; most of them invented or else improved by himself. He composed a catalogue of seven hundred and seventy-seven stars, with greater exactness than had ever been done before; and constructed tables for finding the place of the most remarkable stars at any given time. He was the first who determined the effect of refraction, whereby we see the sun or any star above the horizon, before it is so in reality; as we see the bottom of a vessel when filled with water, standing at a distance, which we could not see when empty. He made several other improvements and important discoveries, which he published in a work entitled "Progymnasmata." The labours of Tycho attracted the attention of Europe; the learned went to consult him, and the noble to see him. James VI. of Scotland, when he went to espouse the sister of Frederic, King of Denmark, paid Tycho a visit, with all his retinue, and wrote some Latin verses in his praise.

But these honours were of short continuance. After the death of his protector, King Frederic, the pension assigned him was withdrawn, and he was compelled to exile himself from his native country. Having hired a ship, he transported his furniture, books, and instruments to a small place in Hamburgh, in 1597. The Emperor Rodolphus invited him into his dominions, settled a large pension upon him, gave him a castle near Prague, to prosecute his discoveries, and appointed him Longomatus, a native of Jutland, and the celebrated Kepler, to assist him. But Tycho was not happy in his new situation; he died 14th October, 1601, repeating several times, "I have not lived in vain."

Kepler was one of the greatest philosophers that ever lived, and ought to be considered as the discoverer of the *true* system of the world. He was born in Germany, at Wiel, near Wirtemberg, 27th December, 1571. He early imbibed the principles of Copernicus. After the death of Tycho, he was employed to finish the tables which he had begun to compose from his observations. Kepler took twenty years to finish them. He dedicated them to the emperor, under the title of the "Rodolphine Tables."

Kepler united optics with astronomy, and thus made the most important discoveries. He was the first who discovered that the *planets move not in a circle, but in an ellipse*; and that altogether they move sometimes faster and sometimes slower, yet that they describe equal areas in equal times; that is, that the spaces through which they move in different parts of their orbit, are of equal times, though of unequal length; yet when two straight lines are drawn from the extremity of either space to the centre of the sun, they form triangles which include equal areas. He likewise demonstrated that the squares of the periodical times of the revolution of the planets round the sun, are in proportion to the cubes of their distance from him; a theorem of the greatest use in astronomical calculations: for having the periodical times of two planets given, and if the distance of one of them be known, by the rule of proportion, the distance of the other can be ascertained.

Kepler is said to have used logarithms in framing his "Rodolphine Tables." This great man died in poverty, 15th November, 1631, at Ratisbon, whither he had gone to solicit the arrears of his pension, which had been very ill paid: he left nothing to his wife and children but the remembrance of his virtues.

Contemporary with Kepler was Galileo, born at Pisa, in Italy, in 1564; illustrious for his improvements in mechanics, for his application of the effects of gravity, and for the invention, or at least, the improvement of telescopes.

The use of spectacles, or reading glasses (convex for long-sighted; and concave for short-sighted persons,) had been invented by one Spina, a monk at Pisa, in 1290; or, as some say, by our countryman Roger Bacon. The use of telescopes or glasses for viewing objects at a distance, was invented by Zachary Janssen, a spectacle-maker, at Middleburg, or rather, as it is said, from the accidental discovery of a child. The honour of this invention is also claimed by others. It is certain that Galileo first improved them so as to answer astronomical purposes. He also first made use of the single pendulum for measuring time in making his observations; to which he was led, by considering one day the vibrations of a lamp suspended from the vaulted roof of a church. He likewise discovered the gravity of the atmosphere from the rising of water in a pump, by the action of a piston, which led the way to the invention of the barometer, by his scholar Toricelli.

The use of telescopes opened, in a manner, a new world to Galileo. He observed with astonishment the increased magnitude and splendour of the planets and their satellites, formerly invisible: which afforded additional proofs of the veracity of the Copernican system, particularly the satellites of Jupiter, and the phasis of Venus. He discovered an innumerable multitude of fixed stars, which the naked eye could not discern, and what greatly excited his wonder, without the least increase in their size or brightness.

About the same time, John Napier, of Merchiston, in Scotland, invented what are called "Logarithms," first published at Edinburgh in 1614, afterwards improved by Mr. Briggs, Professor of Geometry, at Oxford, in which, by a very ingenious contrivance, addition is made to answer for multiplication, and subtraction for division; an invention of the greatest utility in astronomical calculations.

Galileo was not afflicted with poverty, but with persecution. At seventy years of age he was called before the Holy Inquisition, for supporting opinions contrary to

Scripture,—and was obliged, on the 11th of June, 1633, formally to abjure them, to avoid being burnt as a heretic. The system of Copernicus had yet gained but few converts; and the bulk of professions and learned men in Europe, attached to the philosophy of Aristotle, supported the old doctrine. Galileo was condemned to prison, and confined to the small city of Arcem, with its territory, where he consoled himself by the study of astronomy. He contrived a method of discovering the longitude by the satellites of Jupiter, which, however, has not been productive of all the advantages he expected. He died in prison, or rather in exile, in 1642.

Although there were a great number of astronomers contemporary with Kepler and Galileo, none made any conspicuous figure. John Bayer, of Augsburg, introduced the Jewish method of marking the stars with letters of the Greek and Latin alphabets; this the Jews use because their law does not permit the use of figures, the produce of fancy.

In 1732, astronomers were very attentive to observe the transit of Venus over the disc of the sun, which Kepler had predicted, as a confirmation of the system of Copernicus. Mercury was observed by Gassendi in France, and some others; but the transit of Venus did not then take place for their inspection.

The transit of Venus was first seen by Jeremiah Horrox, of Hoole, an obscure village, fifteen miles north of Liverpool, on the 24th of November, 1639, and at the same time by his friend, William Crabtree, at Manchester. Horrox was born in 1619, and died in 1641, in the twenty-third year of his age. He wrote an account of his observations, which were published after his death, under the title of "Venus in Sole visa," by Hevelius.

The Copernican system was first publicly defended in England, by Dr. Wilkins, in 1660; in France, by Gassendi, the son of a peasant in Provence, who published many valuable works on Philosophy. He was born in 1592, and died in 1655. He was violently opposed by Morin, a famous astrologer.

Descartes, descended from a noble family, the son of a counsellor of Brittany, in France, born at Haye, in Tourraine, 31st of March, 1596, early distinguished himself by his knowledge in algebra and geometry. He attacked and overturned the philosophy of Aristotle, in his own country. He attempted to establish certain principles, which, though founded in theory, he took for granted, by which he accounted for all appearances. Like Mochus and Democritus, he imagined all space to be filled with corpuscules, or atoms, in continual agitation, and denied the possibility of a vacuum. He explained everything by supposing vortices, or motions round a centre, according to the opinions of Democritus, and thus discovered the centrifugal force in the circular motion of the planets. But the system of Descartes not being founded on facts or experiments, did not subsist long: although at first it had many followers. His astronomical opinions were much the same with those of Copernicus.

Although the lively notions of Descartes led him into error, yet his exalted views greatly contributed to the improvement of science. Men were led to observation and experiments, in order to overturn his system, and astronomy was cultivated by persons of ability; viz., Bouillard, at Paris; Ward, at Oxford, 1653; and by Helvelius, at Dantzic, 1643, who constructed a fine observatory, and collected a great many facts by his long assiduous observation, for fifty years, during which he made many discoveries concerning the planets, fixed stars, and particularly comets. Colbert, in the name of Louis XIV., sent him a sum of money and a pension. Hevelius published a catalogue of fixed stars, entitled, "Firmamentum Sobieskianum," dedicated to John Sobieski, King of Poland, at that time justly famous for having raised the siege of Vienna, when attacked by the Turks, 1683. In honour of whom Helvelius formed a new constellation between Antinonus and Serpenterius, called *Sobieski's Shield.*

But the most distinguished astronomer of that time was Christian Huygens, son to the secretary of the Prince of Orange, born at the Hague, 14th of April, 1629, and educated at Leyden, under Schooten, the commentator on Descartes,—famous for the application of pendulums to clocks and springs to watches, for the improvement of telescopes and microscopes, and for the great discoveries he made, in consequence of these improvements in astronomy.

The establishment of academies, or societies, at this time, contributed greatly to the advancement of science.

The Royal Society, in London, was begun in 1659, but did not assume a regular form till 1662. Its transactions were first published in 1665. The Academy of Sciences, at Paris, was founded in 1686, by Louis XIV., who invited to it Rœmer, from Denmark, Huygens and Cassini from Italy.

Cassini was born at Perinaldo, in the county of Nice, on the 8th of June, 1625, and was appointed first professor in the Royal Observatory at Paris, where he prosecuted his discoveries till his death, in 1712, and was succeeded by his son. He was assisted by Picard, Auzoul, and La Hire.

By the direction of the Academy of Sciences at Paris, a voyage was undertaken by Riecher and Meurisse, at the king's expense, to the island of Caienne, in South

America, almost under the equator, in 1672, to ascertain several philosophical facts;—the refraction of light, the parallax of Mars, and of the Sun, the distance of the tropics, the variation in the motion of the pendulum, &c.

The parallax of the sun is the angle under which an observer at the sun would see the earth: this Cassini fixed at 9½ seconds, and the angle under which we see the sun, at 16 minutes and 6 seconds, or 966 seconds; hence he concluded that these semi-diameters, are as 9½ to 966, or as 10 to 1932. So that, according to Cassini, the semi-diameter of the earth is one hundred times less than that of the sun; and consequently the sun is a million times larger than the earth.

The parallax of the sun has since, from the transit of Venus, 6th of June, 1761, and 3rd of June, 1769, been discovered to be but about 8 seconds, consequently his comparative bulk to that of the earth, and his distance from it, to be proportionably greater. The method of finding the distance of the earth from the sun, and consequently of the other planets, was first proposed by Dr. Halley, who had never seen, and was morally certain he would never see, this appearance.

Meurisse died during the voyage. Riecher returned in 1676. His answer to the parallax of Mars was not satisfactory. Cassini calculated it at 15 seconds.

The distance of the tropics was found to be 46 degrees, 57 minutes, 4 seconds. The chief advantage resulting from the voyage was ascertaining the vibration of the pendulum. In 1669, Placard remarked that clocks went slower in summer than in winter, owing to, as since ascertained, that it is the property of heat to dilate bodies, which consequently lengthens the pendulum; that cold produces an opposite effect. Riecher found that the pendulum made forty-eight vibrations less at Caienne than at Paris; that it went two minutes and twenty seconds a day slower; hence, to adjust, he was obliged to shorten the pendulum.

The same fact was confirmed by Halley, while at St. Helena, 1676. But an additional reason for this variation is presumed to exist, from the machinery being further removed from the central axis of the earth; the gravitating principle is presumed to be diminished at the equator more than it is when nearer the poles.

About this time the French Jesuit missionaries, having got admission into China, contributed greatly to the improvement of their astronomy. Father Schaal, one of their number, on account of his merit, and particularly for his skill in astronomy, was so highly honoured by the court of China, that the emperor, upon his death-bed, made him preceptor to his son and successor. Schaal reformed the Kalendar, a matter of great importance to that country. It was further improved by Verbiest, who succeeded Schaal, about 1670. The most eminent astronomers in England during this period were Flamstead, Halley, and Hook.

Sir Isaac Newton was born at Woolstrope, in Lincoln, December 25, 1642; after due preparation he was admitted in the University of Cambridge. The rapidity of his progress in mathematical knowledge was truly astonishing. At the age of twenty-four, he had laid the foundation of the most important mathematical discoveries. He is the first who gave a rational and complete account of the laws which regulate planetary motion, on the principles of attraction and gravitation. Newton was as remarkable for a modest diffidence of his own abilities, as for the superiority of his genius. In 1704, he published his "Optics;" in 1711, his "Fluxions;" and in 1728, his "Chronology." He received in his life time the honour due to his singular merit. In 1703, he was elected President of the Royal Society. In 1705, he received the honour of knighthood by Queen Anne.—He was twice member of parliament. In 1669, he was made master of the mint, which, with the presidency of the Royal Society, he held till his death, in 1726. He was buried in Westminster Abbey, where there is an appropriate monument to his memory.

The system of Newton had an eminent supporter and able annotator in the very eminent Scottish professor, Colin Mac Laurin, who was born in the month of February, 1698. In 1719, he travelled to London, where he was introduced to the illustrious Newton, whose notice and friendship he obtained, and ever after reckoned as the greatest honour and happiness of his life. In 1734, Dr. Berkeley, Bishop of Cloyne, published his treatise, called "The Analyst," in which he attempted to charge mathematicians with infidelity in matters of religion. This work was the occasion of Mac Laurin's elaborate "Treatise on Fluxions," published at Edinburgh, in 1742, which is reckoned the most ample treatise on that branch of novel mathematics which has yet appeared. So very eminent was Mac Laurin's skill in mathematics, and the principles of anatomical science, and he possessed such excellent instruments for these purposes, that a new theory never appeared, nor did anything transpire in the scientific world, which was uncommon, but his friends constantly resorted to him for explanation and experiment, and their laudable curiosity was sure to be satisfactorily gratified.

One of the greatest names in the modern history of astronomical discovery is that of the late Sir William Herschel; and, much to his praise, he was self-instructed in the science in which he earned his high reputation. Herschel was born at Hanover, in 1736, and was the son of a musician in humble circumstances. Brought up to his father's profession, at the age of fourteen he was placed in the band of the Hanoverian Guards. A detachment of this regiment having been ordered to England in the year 1757, he and his father accompanied it; but the latter returned to Germany in a few months, and left his son to try his fortune in London. For a long time he had many difficulties to contend with, and he passed several years principally in giving lessons in music in the different towns in the North of England. At last, in 1765, through the interest of a gentleman to whom his merits had become known, he obtained the situation of organist at Halifax; and next year, having gone to fulfil a short engagement at Bath, he gave so much satisfaction by his performances, that he was appointed to the same office in the Octagon Chapel of that city, upon which he went to reside there. The place which he now held was of some value; and from the opportunities which he enjoyed of adding to its emoluments, by engagements at the rooms and private concerts, as well as by taking pupils, he had had the prospect of deriving a good income from his profession, if he had made that his only or his chief object.

During his residence at Bath, although greatly occupied with professional engagements, the time he devoted to his mathematical studies was surprising. Often, we are told, after a fatiguing day's work of fourteen or sixteen hours among his pupils, he would, on returning home at night, repair for relaxation to what many would deem these severer exercises. In this manner, in the course of time, he attained a competent knowledge of geometry, and found himself in a condition to proceed to the study of the different branches of physical science which depend upon the mathematics. Among the first of the latter that attracted his attention, were the kindred departments of astronomy and optics. Having applied himself to these sciences, he became desirous of beholding with his own eyes those wonders of the heavens of which he had read so much, and for that purpose he borrowed from an acquaintance a two-feet Gregorian telescope. This instrument interested him so greatly, that he determined to procure one of his own, and commissioned a friend in London to purchase one for him, of a somewhat larger size. But he found the price was beyond what he could afford. To make up for this disappointment, he resolved to construct a telescope for himself; and after encountering innumerable difficulties in the progress of his task, he at last succeeded, in the year 1774, in completing a five-feet Newtonian reflector. This was the commencement of a long and brilliant course of triumphs in the same walk of art, and also in that of astronomical discovery. Herschel now became so much more ardently attached to his philosophical pursuits, that, regardless of the sacrifice of emolument he was making, he begun gradually to limit his professional engagements, and the number of his pupils.

Meanwhile he continued to employ his leisure in the fabrication of still more powerful instruments than the one he had first constructed; and in no long time he produced telescopes of seven, ten, and even twenty feet focal distance. In fashioning the mirrors for these instruments, his perseverance was indefatigable. For his seven-feet reflector, we have been informed that he actually finished and made trial of no fewer than two hundred mirrors before he found one that satisfied him. When he sat down to prepare a mirror, his practice was to work at it for twelve or fourteen hours, without quitting his occupation for a moment. He would not even take his hand from what he was about, to help himself to food; and the little he ate on such occasions was put into his mouth by his sister. He gave the mirror a proper shape, more by a certain natural tact than by rule; and when his hand was once in, as the phrase is, he was afraid that the perfection of the finish might be impaired by the least intermission of his labours.

It was on the 13th of March, 1781, that Herschel made the discovery to which he owes, perhaps, most of his reputation. He had been engaged for nearly a year and a half in making a survey of the heavens, when, on the evening of the day that has been mentioned, having turned his reflector (an excellent seven feet reflector of his own constructing) to a particular part of the sky, he observed among the other stars one which seemed to shine with a more steady radiance than those around it; and on account of that and other peculiarities in its appearance, which excited his suspicions, he determined to observe it more narrowly. On reverting to it after some hours, he was a good deal surprised to find that it had perceptibly changed its place—a fact which the next day became more indisputable. At first he was somewhat in doubt whether or not it was the same star which he had seen on these different occasions; but, after continuing his observations for a few days longer, all uncertainty upon that head vanished. He now communicated what he had observed to the astronomer royal, who concluded the luminary could be nothing else than a new comet. Continued observation of it, however, for a few months, dissipated this error; and it became evident that it was in reality a hitherto undiscovered planet. This new world so unexpectedly found to form a part of the system to which our own belongs, received from Herschel, the name of the *Georgium Sidus*, or Georgian Star, in honour of the King of England; but by continental astronomers it has been more generally called either *Herschel*, after its discoverer, or *Uranus*. Subsequent observations, made chiefly by Herschel himself, have ascertained many particulars regarding it, some of which are well calculated to fill us with astonishment at the powers of the sublime science which can wing its way so far into the immensity of space, and bring us back information so precise and various. In the first place, the diameter of this new globe has been found to be nearly four and a half times larger than that of our own. Its size altogether is about eighty times that of our earth. Its year is as long as eighty-three of ours.

Its distance from the sun is nearly eighteen hundred millions of miles, or more than nineteen times that of the earth. Its density, as compared with that of the earth, is nearly as twenty-two to one hundred; so that its entire weight is more than eighteen times that of our planet. Finally the force of gravitation near its surface is such, that falling bodies descend only through fourteen feet during the first second, instead of thirty-two feet as with us. Herschel afterwards discovered no fewer than six satellites, or moons, belonging to his new planet.

The announcement of the discovery of the Georgium Sidus at once made Herschel's name universally known. In the course of a few months the king bestowed on him a pension of three hundred pounds a year, that he might be able entirely to relinquish his engagements at Bath; and upon this he came to reside at Slough, near Windsor. He now devoted himself entirely to science; and the construction of telescopes, and observations of the heavens, continued to form the occupations of the remainder of his life. Astronomy is indebted to him for many other most interesting discoveries besides the celebrated one of which we have just given an account, as well as a variety of speculations of the most ingenious, original, and profound character. But of these we cannot here attempt any detail. He also introduced some important improvements into the construction of the reflecting telescope—beside continuing to fabricate that instrument of dimensions greatly exceeding any that had been formerly attempted, with the powers surpassing in nearly a corresponding degree, what had ever been before obtained.

The largest telescope which he ever made, was his famous one of forty feet long, which he erected at Slough for the king. It was begun about the end of the year 1785, and on the 28th of August, 1789, the enormous tube was poised on the complicated but ingeniously contrived mechanism by which its movements were to be regulated, and ready for use. On the same day a new satellite of Saturn was detected by it, being the sixth which had been observed attendant upon that planet. A seventh was afterwards discovered by means of the same instrument. This telescope has been taken down and replaced by another of only half the length, constructed by Mr. J. Herschel, the distinguished son of the subject of our present sketch. Herschel himself eventually became convinced that no telescope could surpass, in magnifying power, one of from twenty to twenty-five feet in length. The French astronomer, Lalande, states that he was informed by George III. himself, that it was at his desire that Herschel was induced to make the telescope at Slough of the extraordinary length he did, his own wish being that it should not be more than thirty feet long.

So extraordinary was the ardour of this great astronomer in the study of his favourite science, that for many years it has been asserted, he never was in bed at any hour during which the stars were visible. And he made almost all his observations, whatever was the season of the year, not under cover, but in his garden, in the open air— and generally without an attendant. There was much that was peculiar to himself, not only in the process by which he fabricated his telescopes, but also in his manner of using them. One of the attendants in the king's observatory at Richmond, who had formerly been a workman in Ramsden's establishment, was forcibly reminded, on seeing Herschel take an observation, of a remark which his old master had made. Having just completed one of his best telescopes, Ramsden, addressing himself to his workman, said, "This, I believe, is the highest degree of perfection we opticians by profession will ever arrive at; if any improvement of importance shall ever after this be introduced in the making of telescopes, it will be by some one who has not been taught by us."

Some years before his death, the degree of LL.D. was conferred upon Herschel by the University of Oxford; and in 1816, the Prince Regent bestowed upon him the Hanoverian and Guelphic Order of Knighthood. He died on the 23rd of August, 1822, when he was within a few months of having completed his eighty-fourth year.

We have been thus particular in the enumeration of particulars in the lives of those great men, who have cultivated this sublime science, for the purpose of availing ourselves of a suggestion furnished by Dr. Priestly, who observed, "That we could only see Newton in two points of his career: at the bottom of the ladder, and at the top; having left no account of his progress, it appeared as though he had broken the steps by which he had ascended, that none should follow."

From the facts collected by the many eminent men whose names have ornamented our pages, we are enabled to state the following particulars concerning that part of the universe denominated the Solar system.

The *Sun*, a luminous body diffusing light and heat; whose diameter is computed at 890,000 miles; diurnal rotation on axis 25 days 6 hours; performs his annual revolution in orbit in 365 days 6 hours; progressive equatorial motion in orbit per hour, 3818 miles.

Mercury, whose diameter is 3,000 miles, revolves in an orbit 36,481,448 miles from that of the sun. He performs his annual period round that planet in 87 days 23 hours; his hourly equatorial motion in orbit is 109,699 miles.

Venus,—her diameter is 9,330 miles; revolves in an orbit 68,891,486 miles distant from the sun; performs her annual revolution in 224 days 17 hours; diurnal rotation on axis 24 days 8 hours: hourly equatorial motion in orbit 80,295 miles.

The *Earth,*—its diameter 7970 miles; distance of orbit from the sun 95,173,000 miles; revolves on its axis once in 24 hours; performs her annual period round the sun in the same time the sun completes his revolution; hourly equatorial and progressive motion in orbit 80,295 miles.

The *Moon* is a satellite to the earth; her diameter is 2180 miles; her diurnal rotation on axis is performed in 29 days, 12 hours, 44 minutes; she performs her annual revolution round the sun in precisely the same time as does the earth, her superior planet; her motion in orbit per hour is 22,290 miles.

Mars,—his diameter is 5400 miles; distance from the sun, 145,014,148 miles; annual period round the sun 671 days, 17 hours; diurnal rotation on axis 19 days, 12 hours, 44 minutes; hourly motion in orbit 55,287 miles.

Jupiter,—his diameter 94,000 miles; distance from the sun 494,990,976 miles; annual period in 11 years, 314 days, 18 hours; diurnal rotation on axis 9 hours, 56 minutes; hourly motion in orbit 29,803 miles.

Saturn,—his diameter is 78,000 miles; distance from the sun 907,956,130 miles; annual revolution in orbit 22 years, 167 days, 6 hours; hourly motion in orbit 22,101 miles.

It should be observed that Jupiter has four moons, or satellites, with a large and very luminous belt at a great distance from his surface. Saturn also has seven moons, with a very luminous ring about 21,000 miles broad, from its uppermost to its undermost edge; and about the same distance from its surface.

Georgium Sidus,—the distance of the orbit from the sun, 1,758,000,000 miles; annual revolution 28 years, 289 days; diameter 56,726 miles; has two satellites, or moons.

About 1801, 2, and 4, there were discovered three other small planets in the system of the sun, called *Vesta*, *Juno*, and *Pallas*.

The fixed stars composing the *Zodiacal Signs*, are divided into twelve constellations, one to each month; which asterisms were discovered by Flamstead to consist of the following number of stars to each:

Aries, the Ram, 66; *Taurus*, the Bull, 141; *Gemini*, the Twins, 85; *Cancer*, the Crab, 83; *Leo*, the Lion, 95; *Virgo*, the Virgin, 110; *Libra*, the Scales, 51; *Scorpio*, the Scorpion, 44; *Sagitarius*, the Archer, 69; *Capricornus*, the Goat, 51; *Aquarius*, the Water-Carrier, 108; *Pisces*, the Fishes, 113.

A comparative idea of the extent of the works of Omnipotence may be perhaps collected, on our being informed, that the sphere where the fixed stars appear, is presumed to be placed far beyond the most remote planetary orbit; and that some of them are supposed to serve as suns to illumine other systems, or worlds, to us unknown.

NAVIGATION.

THE sacred records inform us that the ark of Noah was the first ship, and produced by the invention of the great Architect of Nature himself; and "though some men have so believed," says the learned and ingenious Sir Walter Raleigh, in his "History of the World," "yet it is certain the world was planted before the flood, which could not be performed without some transporting vessels. It is true, and the success has proved that there was not any so capacious, nor any so strong, as to defend themselves against so violent and so continued a pouring down of rain, as the ark of which Noah was the builder, from the invention of God himself. Of what fashion or fabric soever were the rest, with all men they perished according to the ordinance of God." And it appears extremely probable that those testimonials, whereof Ovid speaks of former existence, were remains of ships wrecked at the general flood.

There can be no question that the Syrians were the first maritime power in the world, as well in point of time as importance;—but of what species of construction their vessels were, we are not informed. Their merchants trading to the Eastern Indies, as they did for Solomon; to Ophir, whence they brought gold; and also to this country for tin, and their having made three distinct descents upon America, will enable us to maintain this our opinion. After them the Greeks, a people living chiefly on the shores of the Hellespont and Ægean seas, with many islands in the Mediterranean, Adriatic, and Archipelagion Seas, besides their possessions in Asia Minor, and their commerce with the European Continent, obtained the next power by sea. We read indeed, that Minos, the famous Cretan sovereign and legislator, who lived two descents before the Trojan war, sent out shipping to free the Grecian seas of pirates; which shows, as Sir Walter Raleigh ingeniously infers, that there had been trade and war upon the waters before his time also.

The next in point of time and importance on record was the highly renowned expedition of the Argonauts for the golden fleece to Colchis, a country of Asia, on the Euxine sea. Immediately after this was the colonization of Cyrene, in Africa, by Battus, one of the companions of Jason, in his Colchian expedition. Shortly afterwards, the Grecian states united against Phrygian treachery and the abuse of Grecian hospitality; forming another most memorable epoch in the history of the world. We are informed the Grecian Neptune, or as mythology styles him a God of the Saturnian family, for the great service he did his father, Saturn, or Noah, against the Titans, had the seas given to him. History informs us that the first inventor of rowing vessels was a citizen of Corinth; and likewise that the first naval war was between the Samians and Corcyrians. The history of Ithicus, translated into Latin by St. Jerome, affirms that Griphon, the Scythian, was the inventor of long-boats; and Strabo also gives the honour of the invention of the anchor to another Scythian, the famous Anacharsis, whilst Greece herself by her historians, ascribes its invention to Eupolemus. Also, it is said, that Icarus invented the sail, and others, various other pieces of the component parts of ships and boats. The specification of such other imperfect memoirs, many of fabulous appearance, may be of no great importance.

It appears certain that among the four sons of Javan, the son of Japhet, the grandson, and other the posterity of Noah, who peopled the "Isles of the Gentiles," the Grecian Islands must long before the days of Minos have used those seas, from the insular nature of their inhabitants. And it certainly does not appear extravagant to us, to presume that this people were among the first who navigated the seas. Mankind in various parts of the world, being stimulated by the same necessities, urged by the same wants, and possessing the same means, might probably produce similar inventions to each other. Most, indeed, had occasion to navigate lakes, and cross rivers. They accordingly constructed such machines as would answer their purpose of passage or migration. So were rafts and canoes, formed of canes, osiers, twigs, &c., where they grew, which they fashioned like boats, and then covered with skins of various animals; others formed rafts of wood; whilst some others fashioned canoes, having hollowed out trees for that purpose. One way or other, each people thus possessed a marine, proper for their purpose it is true, but in various degrees of excellence. This was the case with Greeks as well as barbarians of all nations; all these people, excepting the immediate descendants of Noah, might, perhaps, lay a feasible claim to the honour of the original invention of these articles; and, having never seen such, they virtually have each a good title to the distinction. Indeed, many of them might have taken the idea for such invention from the policy of certain animals, and the nature of others; to instance the sagacity of the beaver and his raft, and the little nautilus with his swelling sail: hence they might have adopted from that animal, and that piscatory insect, the idea of a raft, and also of a vessel with a sail.

In latter days we find the Teutonic Saxons first came to this country, according to Mr. Turner, the Anglo-Saxon historian, in vessels they called *cyules-kells* by Sir Walter Raleigh. Marine vessels have borne a variety of names, as well as of numerous figures, from the gondola of the Venetian to the canoe of the Esquimaux,—the British man-of-war to the ponderous bonaventure in which the Doge annually espouses the sea.

All those nations, too, through whose hands the maritime power has passed, from time to time, as they have been instructed by experience, or taught by necessity, might repeatedly have made additions and improvements in naval architecture: some calculated for mercantile utility, while others have only attended to warlike strength, and some to answer both purposes, like our Indiamen. But now, the British navy, being supplied with the best materials, and having as ingenious workmen as any, with the addition of the warlike children of the soil, may openly defy all nations, and proudly claim the sovereignty of the seas where her flag has been flying 'midst the battle and the breeze for so many years.

But the most important improvement in Navigation— propelling vessels by steam—has been left to our own times. The steam-engine was first applied to small vessels for the coasting or river trade; but it has now increased to vessels of the largest size,—in fact, the most part of the British navy are steamships. In former times before the introduction of this valuable auxiliary, the passage between England and America was tedious and uncertain, sometimes taking months, but rarely less than from four to six weeks, according to the state of the weather; but now the case is altered. There are a regular line of steamships, one of which leaves Liverpool every week, and the voyage is performed with almost positive certainty in from twelve to fourteen days, independent of the rude Boreas, or the boisterous Atlantic. These vessels are of the largest size and handsomely fitted up for the accommodation of passengers.

LIGHT-HOUSES.

A light-house, in marine architecture, is a building, or watch-tower, erected on the sea-shore, to serve as a land-mark to mariners, on a low coast, by day, and, in any situation, to inform them of their approach to land in the night;—being of most essential utility in causing them to take soundings, avoid shoals, rocks, &c.; or else it is a building erected on a rock in the sea, which, from its situation, would be extremely dangerous to vessels, were not some intimation given of the existence of a rock, where it is locally situated. Of this latter description is the celebrated Eddystone light-house, off Plymouth.

Although this species of architecture is not likely to have been so general in extreme antiquity, because it could not have been essentially necessary to any except to those nations who, from the proximity of their situation to the coast, or other circumstances, pursued maritime concerns; or to those whose connexions rendered the encouragement of the marine of other nations important.

The oldest building of this description, which we believe to be upon record, is the famous Pharos erected on the Egyptian coast, which, being very low land, and exposed entirely to the almost constant west winds coming up the Mediterranean from the vast Atlantic, must, of necessity, have made the port of modern Alexandria, anciently called Dalmietta, very dangerous. It was originally erected by Ptolemy Philadelphus, for the encouragement and convenience of the Phœnicians, who were accounted the foreign factors of that empire; as the Egyptians possessed an unconquerable aversion to the sea, and therefore they never obtained its sovereignty: whilst the former people were the first who obtained the supremacy of that sea.

The island upon which Pharos stood, in the time of Homer, in his simple geography and estimation, was said to be one day's sail from the Delta; whereas, since the foundation of Alexandria, it was only a mile in distance, and was even joined to the mainland by a mole, having a bridge at each end; or according to some authors, in the middle. The tower was, if report be true, justly entitled to the appellation it obtained—one of the seven wonders of the world; and it is reported, that the light from it has been seen at the distance of a hundred miles; which, assuredly, appears improbable, because the convexity of the earth, we think, would not permit. Its height must have been, at least, 2,400 feet, or 800 yards from the base.

We are enabled to furnish the following particulars of this famous structure. It was built by order of that patron of learning and the arts, Ptolemy Philadelphus, by that eminent architect, Sostrates, who constructed many of the public buildings in Alexandria. It is said to have cost Ptolemy eight hundred talents! Respecting its mode of construction, it was raised several stories one above another; each was decorated with columns, balustrades, and galleries of the finest marble and most exquisite workmanship; and some have even said that the architect had furnished the galleries with large mirrors, by which shipping could be seen at a great distance. However, respecting this edifice, once so famous, that its very name, Pharos, was considered as a common term for all other constructions for the same purpose, it is now said, from Saracenic ignorance and brutality, aided, perhaps, by the assistance of the common leveller, Time, that nothing now remains of this once elegant edifice, but an unsightly tower rising out of a heap of ruins, the whole being accommodated to the inequality of the ground on which it stands, and being, at present, no higher than that which it should command. Such as it is, there is now a light, we understand, usually maintained. There is also an island, which was called Pharos, in the Adriatic sea, on the coast of Italy, opposite Brundusium, for the same reason: likewise the celebrated colossal statue of Apollo, at Rhodes, answered the same purpose, and occasionally had the same appellation, as had a river of Asia, in the environs of Cilicia and the Euphrates. This last consideration brings us to the etymology of the word, as Ozanum says, "Pharos originally signified a strait, as the Pharos of Messina." Of every description of light-houses yet known, there is none more famous than that called Eddystone, with a description of which we shall conclude this article.

Mr. Winstanley's light-house was begun upon the Eddystone rock in 1696, and was more than four years in building, from the numerous interruptions of the wind and the element he had to contend with, the violence whereof is truly alarming, occasioned by that rock being exposed to every wind which comes up the vast Atlantic, and that tumultuous sea, the Bay of Biscay. These obstacles were considerably increased by the shape of the rock itself, having a regular slope to S.W., and from the very deep sea in its vicinity, it, therefore, receives the uncontrolled fury of those seas: meeting with no other object whereon to break their vehement force, the effect is so great at high water with a S.W. wind, which continues for many days, though a calm may have succeeded, the violent action of the waters has not ceased, but break frightfully on Eddystone. An engraving of Mr. Winstanley's light-house was published at the period of its erection, from which it appears to have been a stone tower of twelve sides, rising forty-four feet above the highest point of the rock, which, in the dimensions on which it was built, twenty-four feet in diameter, was ten feet lower on one side than it was upon the other; at the top was a balustrade and platform; upon this were erected eight pillars, which supported a dome of the same dimensions as the tower; from the top of which arose an octagon tower, of a diameter of fifteen feet, and seven in height. On the summit was placed the lantern, ten feet in diameter, and twelve in height: it had a gallery surrounding it, which gave access to the windows. The whole was surrounded by fencible iron-work. The entry was by a solid stone door at the bottom; the whole building was of the same material, except the aperture for the staircase. At the bottom was a room twelve feet high for a store-room; the next story was of the same height, which was the stateroom; and the third was of a similar height, which was the kitchen. Those compartments occupied the whole height to the platform. The dome above this contained the lodging-room; the octagon above it, the look-out.

The reason why it occupied so much time in building was, because the men could only work in the summer months. The first summer was occupied in making holes in the rock, and fastening irons to hold the future work. The second year was spent in erecting a solid pillar, of fourteen feet diameter, and one hundred and twelve feet high, for the future support of the building. The third year, it was augmented in diameter and increased in height. This building was eventually finished, within the time above-mentioned, at an enormous expense. It stood the opposition of the elements. The violence of the sea was so great, that Mr. Winstanley said it has been seen to rise upwards of one hundred feet above the vane, whilst the sides of the building were covered with surf as with a sheet, so that the whole house and lantern were occasionally under water. This edifice withstood the conflict of elements till 1703, when the architect, being at

Plymouth, and desirous of visiting it, for the purpose of inspecting some repairs, went to it, but returned no more; for a storm arose, which left not a relic of it standing, except the iron work, which had been fixed in the rock. The Corporation of the Trinity House had then to erect another, for which purpose they employed a Mr. John Rudyard, who was a silk mercer, on Ludgate-hill. Mr. Rudyard's mechanical ingenuity was said to have qualified him well for the undertaking. It appears that he erected a house made chiefly of wood, which presented many traits of his genius. It was a conical frustrum, one hundred and fifty-six feet in diameter at the base; its altitude sixty-two feet. At the top of the building was a balcony, railed round; in the centre of its area was the lantern. This building was made quite plain, excepting the well for the staircase, which was solid for thirty-two feet. In the centre a strong mast was erected. The building was admirably fixed to the rock, from the very peculiar manner of making the holes to hold iron cramps, they being made for the internal cavity to diverge on each side, by an extreme of one inch at the depth of sixteen inches. The cavity was first filled with tallow; the hot iron then dipped in the same substance, put in the rock, and eventually filled with pewter, which displaced the tallow, being heavier, the grease serving to protect the iron from the corrosive acidity of the salt water. In 1708, it was finished so far as to receive a temporary light. It stood forty-four years, and showed that it was liable to destruction from the very perishable nature of its materials. However, on the 2nd December, 1755, the upper part of it taking fire, burnt downwards to its entire consumption. The concern had been leased to a Captain Lovell; but at a later period his possessions were distributed among a number of people, when the care of rebuilding it was entrusted to Mr. Robert Weston, to whom Mr. John Smeaton was recommended by the President of the Royal Society, who appears to have been well qualified for the undertaking. He accordingly furnished a plan for, and superintended the building which now stands. Mr. Smeaton's conjecture was quite different to that of the late projector; he conceived that nothing could withstand the action of the wind and water so well, and at the same time, prevent such accidents as the past, as could a building whose gravity should secure its most sure protection, He accordingly constructed his of the most massy stones, all dovetailed into each other, formed of Cornish-moor and Portland stone; all the joints breach each other, as the masons term it, or on each joint occurs the central stone of the next course. There are fourteen courses of these stones first laid in this manner, of a great thickness each course. On the 12th June, 1757, the first stone was laid in its place, each stone being pierced when it was laid, a strong oak pin was driven through to pin it fast to its place: the dovetails not fitting so close to each other, because it was necessary to leave some space for the cement, this pin was calculated to secure the stone till this could be applied and had fixed; the cement used was composed of Watchet lime and *puzzolana*, or Dutch terras, being made at the moment by mixing up in a pail, with water; this mixture was poured upon the work, and run into every cavity and crevice; this, however, was sometimes not exempt from the injury of the sea; whenever it was injured, the defect was supplied by having some oakum cut fine, and mixed with this cement, introduced into the joints; then they were secured with a coat of plaster of Paris, *pro tempore*, and this was never known to fail, if the work stood for one tide. In this manner the platform was erected, all of the most solid materials, and substantial workmanship.

On the 30th of September, 1758, the work having been continued from the 11th of the preceding May, had arrived at the store-room floor; here an iron chain was let into the stone, as follows: the recess being made and the chain being well oiled before insertion, the groove which received it was divided into four separate dams by clay; two kettles were used, to hold a sufficiency of melted lead, eleven hundred weight; whilst the lead was in a state of fusion, two men with ladles filled one quarter of the groove; as soon as it set, they removed one of the clay dams, and then filled the next quarter, pouring the liquid on the middle of the first quarter, it melted together into the second; the dam at the opposite end was now filled, and then the fourth; by this means the lead was associated into one solid mass. The centring for the floor was next set up, the outward stones being first set, and then the inner ones. Thus the base floor was finished. The men could work no longer than till the 7th of October that year. The winter was spent in preparing the iron, copper, and glass work for the lantern; and the spring in unsuccessful endeavours to discover the moorings for the vessel which attended the works, for the occasional retreat of the workmen. On the 5th of July the work was resumed: the stones for building had been hitherto raised from the boats by what are called shears, formed of two poles, with the lowermost ends extended to a sufficient width, whilst the upper ends met in a point; here was fastened tackle, pulleys, &c., to raise them to a sufficient height to be swung over the building; this course was now of necessity altered; a block with pulleys being suspended from the top, projected to a sufficient distance, supported by beams. After the base had been formed as described, a different mode of operation was necessary to complete the superstructure; the work being now advanced so high as to be out of the constant wash of the sea. Instead of grooves being formed to fasten the stones together, they were fixed by means of iron clamps and lead. The stones to complete the superstructure were landed, and first drawn up by machinery, called a *jack*, through the well, in the interior of the building, being a cavity for the staircase. The work now proceeded more rapidly, so that by the 26th of August, the stairs and all the masonry were finished: the iron frame for the lantern was next screwed together in its place, and the lantern soon completed. It should have been noticed, that after the first entry was closed, the shears were supported by a tackle called a *guy*, attached to the top of the shears, and hooked so far on the outside of the building; the stone being drawn up by a windlass, the guy was drawn in to swing the stone over the building. The balcony rails and the stone basement for the lantern having been completed, on the 17th of September the cupola was set up by a particular kind of shears constructed purposely, the guy in different places being fastened to booms projecting from the several windows of the upper rooms; the next day the ball was screwed on, and on the 11th of October, an electrical conductor was fixed, which finished the edifice. A light was then exhibited, which has continued to warn the mariner ever since. An ably constructed cornice throws the spray from off the building, so that it is often seen at Plymouth with the appearance of a white sheet, throwing itself to double the height of the building, which from low water mark to the apex of the ball is one hundred feet.

We have been thus minute, because this pharos is considered to be the best constructed of all our lighthouses.

ELECTRICITY.

ELECTRICITY was a property but imperfectly understood by the ancients; indeed, it has been said, they were entirely unacquainted with it. But we propose, shortly, to show the extent to which we are informed their sphere of knowledge extended. This much cannot be denied, that they were acquainted with the electrical properties of amber, of which fact we are informed by Pliny.

Even before Pliny, however, as early as the days of Thalis, who lived near six hundred years anterior to the Roman historian, the Miletine philosophers ascribed the attractive power of the magnet and of amber to animation by a vital principle. Our word "electricity" appears to be derived from the name the Latins gave to amber, *electrum*. It is also evident that they were acquainted with the shock of the torpedo; although they were ignorant, as are the moderns, of the concealed cause of this effect.

It has been asserted that the ancients knew how to collect the electrical fire in the atmosphere; and it is also said, that it was in an experiment of this nature that Tullus Hostilius lost his life. Etymologists have carried us still farther back, and assert that it was from the electrical property in the heavens that Jove obtained his surname of Jupiter *Eliaus*. This, however, may be only conjectural.

The first discoveries made of sufficient importance to demand the appellation of "scientific" in the science of electricity, were effected by Dr. W. Gilbert, the result of which he gave the world, in the year 1660, in a book then published, entitled "De Magneto," and Dr. Gilbert was followed in his pursuits by that celebrated scientific character, the honourable and illustrious Boyle, and other men eminent for that species of information.

This science was successfully cultivated in the last century by many eminent philosophers, among whom we may mention Hawkesbee, Grey, Muschenbrook, Doctors Franklin and Priestly, Bishop Watson, Mr. Cavendish, and several other members of the Royal Society of England; whilst those worthy of the true philosophic character in France did not neglect its cultivation.

Many fatal accidents have resulted from experiments made by people ignorant of the science. On the 6th of August, 1753, at Petersburg, Professor Richmann lost his life by endeavouring to draw the electric fluid into his house.

Electricity, like many others of the arcana of nature, still retains almost as deeply shaded from human view as when its existence was first made known. Nature appears to have certain secret operations, which are not yet, perhaps, to be revealed.

ELECTRIC TELEGRAPH.

This is the most surprising invention of modern times, and of the greatest importance to a commercial people; by means of it intelligence is conveyed from one end of the kingdom to another, in the twinkling of an eye. A company was fully organised for the carrying out this invention, which commenced its operations in 1848, and established a system of no ordinary complication and extent. Their wires stretch from Glasgow on the north, to Dorchester, on the south, from the east coast, at Yarmouth, to the west, at Liverpool. These have brought upwards of one hundred and fifty towns into instant communication with each other. The wires set up for the use of the public alone are upwards of nine thousand eight hundred miles in length, and extend over a distance of two thousand and sixty miles, and, exclusive of those running underground, and through tunnels or rivers, are stretched on no fewer than sixty-one thousand eight hundred posts, varying from sixteen to thirty feet in height, and of an average square of eight inches, with an expensive apparatus of insulators and winders attached to each. As the most trifling derangement of the wires or apparatus will stop the communication, it is obvious that the utmost care and watchfulness is requisite to prevent and detect accidents. Accordingly, the whole distance is divided into districts, each district having a superintendent, and under him several inspectors, and a staff of workmen, batterymen, and mechanics, more or less numerous, according to the extent over which he presides.—When we consider these things, in conjunction with the central staff of engineers, secretaries, &c., at the head-establishment in London, a maximum charge of one penny per mile cannot be considered an exorbitant demand for the accommodation afforded to the public in keeping open so many receiving stations, and the maintenance of the expensive establishments. The telegraphic system is designed for important and urgent messages, and it may be safely averred that not one despatch in a hundred has been as yet forwarded by it, which has not been by many times worth more than the sum paid by the sender. A commercial house in Liverpool will scarcely grudge 8s. 6d. for a communication by which a necessary payment may be made, an important order given, or a profitable operation facilitated in London; and the message from Glasgow, which traverses a distance of five hundred and twenty miles in an instant, to summon a son from the metropolis, it may be, to the bedside of a dying parent, cannot be judged exorbitant at a charge of 14s., considerably less than one halfpenny per mile.

Messrs. Wilmer and Smith, of Liverpool, publishers of the "European Times," have arranged the most admirable code of signals in the world; and by the use of forty-eight letters are capable of transmitting intelligence equal to half a column of an ordinary newspaper. The telegraphic company disapprove of this species of short-hand, and, therefore, charge for the forty-eight letters 13s. This Messrs. Wilmer and Smith consider excessive, as they have forwarded similar messages by telegraph, four thousand miles in America, for 8s., and from Philadelphia to New York for 1s. These gentlemen, therefore, consider they have cause to find fault with the company in reference to charges for communications in cipher.

STEAM-ENGINES.

THE Steam-Engine is one of the most important of human discoveries, and is certainly one of those which afford the greatest portion of ease and advantage to the human species, as well in the operation of its cause, as in its ultimate effects. The most powerful of machines had its origin from the single idea of one individual of our own nation. It has been, from time to time, improved by different individuals, also natives of Britain, the precise period of which improvements can be traced, and their effects fortunately ascertained.

Although we should observe, that the first principle of this mechanical power was discovered by some of the ancient nations, many ages before that which gave the origin to the present practised invention, but from the state of information, it is conceived, to answer no purpose of utility. It may be said to have occurred in a small machine which the ancients called an *Æolipila* (the bull of Æolus) consisting of a hollow ball of metal, with a slender neck, or pipe, also of metal, having a small orifice entering into the ball, by means of a screw; this pipe being taken out, the ball being filled with water, and the pipe again screwed in, the ball is heated—there issues from the orifice, when sufficiently hot, a vapour, with great violence and noise; care was required that this should not be by accident stopped, if it were, the machine would infallibly burst, and perhaps, to the danger of the lives of all in its vicinity, so immense is its power.

Another way of introducing the water was first to heat the ball when empty, and then suddenly to immerse it in water. Descartes, in particular, has used this instrument to account for the natural generation of winds. Chauvin thinks it might be employed instead of bellows, to blow a fire. It would admirably serve to fumigate a room, being filled with perfume instead of common water. It is said to have been applied to clear chimneys of their soot, a practice still alleged to be common in Italy. Dr. Plott, in his "History of Staffordshire," records this singular custom, where the Æolipila is used to blow the fire. "The lord of the mannor of Essington is bound by his tenure to drive a goose, every New Year's day, three times round the hall of the Lord of Hilton, while Jack of Hilton, a brazen Æolipila, blows the fire." The last circumstance we shall mention of this instrument, has relation to an antique one, discovered whilst digging the Basingstoke canal, representing a grotesque metallic figure, in which the blast proceeded from the mouth. This figure is now in the possession of the Society of Antiquaries of London. In this instrument, the uncommon elastic force of steam was recognised before the suggestion of the Marquis of Worcester, which follows:

"In 1655, or subsequent thereto, the Marquis of Worcester published the earliest account of the application of this power for the purposes of utility, and suggested it as applicable to raising water. 'Sixty-eight. An admirable and most forcible way to drive up water by fire; not by drawing or sucking it upwards, for that would be what the philosopher calleth it, *intra spherum actroctatis*, which is, but at such a distance. But this way has no bounder, if the vessel be strong enough; for I have taken a whole piece of cannon, whereof the end was burst, stopping and screwing up the broken end, as also the touch-hole; and making a constant fire under it, within twenty-four hours it burst and made a great crack: so that having a way to make my vessels, so that they are strengthened by the force within them, and the one to fill after the other, I have seen the water run like a constant fountain stream, forty feet high;

one vessel of cold water being consumed, another begins to force and refill with cold water, and so successively; the fire being tended and kept constant, which the self-same person may likewise abundantly perform, in the interim between the necessity of turning the cocks.'"

The marquis's ingenuity did not, it appears, meet with that attention which it deserved, from those to whom his communication was addressed. In the article of steam it has been since very much improved, and is acted upon for the most useful of purposes; also his ideas for short-hand telegraphs, floating baths, escutcheons for locks, moulds for candles, and a mode to disengage horses from a carriage, after they have taken fright; which, with several others, proclaim the originality and ingenuity of the mind of this nobleman—an honour which very few of the British nobility aspire to.

Since his time, another design upon the same principle has been projected by Captain Thomas Savery, a commissioner of sick and wounded, who in the year 1691 obtained a patent for "a new invention for raising water, and occasioning motion to all sorts of mill-work, by the impellant force of fire." This patent bears date the 25th of July, sixteenth of William III., A. D. 1698. The patent states that the invention will be of great use for drawing of mines, serving towns with water, and working all sorts of mills. "Mr. Savery, June 14th, 1699, entertained the Royal Society with showing a model of his engine for raising water by help of fire, which he set to work before them; the experiment succeeded according to expectation."

The above memoir is accompanied with a copperplate figure, with references by way of description; from whence it appears, that the engine then shown by Captain Savery was for raising water, not only by the expansive force of steam, like the Marquis of Worcester's, but also by the condensation of steam, the water being raised by the pressure of a rarified atmosphere to a given height, by the expansive force of steam, in the same manner as the Marquis proposed. This action was performed alternately in two receivers, so that while the vacuum formed in one was drawing up water from the well, the pressure of steam in the other was forcing up water into the reservoir; but both receivers being supplied by one suction-pipe and one forcing-pipe, the engine could be made to keep a continual stream, so as to suffer very little interruption. This engine of Captain Savery's displays much ingenuity, and is almost as perfect in its contrivance as the same engine has been made since his time. We regret, that without a figure we cannot supply a perfect description of it.

However, it appears that it was necessary to have two boilers, or vessels of copper, one large and the other smaller: those boilers have a gauge-pipe inserted into the smaller boiler, within about eight inches of its bottom, and about the centre of the side of the larger boiler; the small boiler must be quite full of water, and the larger one only about two-thirds full. The fire is then to be lighted beneath the larger boiler, to make the water boil, by which means the steam being confined, will be greatly compressed, and will, therefore, on opening a way for it to issue out (which is done by pushing the handle of a regulator from the operator), rush with great violence through a steam-pipe into a receiver, driving out all the air before it, sending it up into a force-pipe through a clack, as may be perceived from its noise; when the air is expelled, the receiver will be very much heated by the steam. When it is thoroughly emptied of atmospheric air, and grown very hot, which may be both seen and felt, then the handle of the regulator is to be drawn towards the operator, by which means the first steam-pipe will be stopped, so that no more steam can rise into the first receiver, by which means a second receiver will be filled in like manner. Whilst this is doing, some cold water must be poured on the first receiver, by which means the steam in it will be cooled, and thereby condensed into smaller room: consequently the pressure in the valve, or cock, at the bottom of the receiver—there being nothing to counterbalance the atmospheric pressure at the surface of the receiver in the inner part of the sucking-pipe, it will be pressed up into the receiver, driving up before it the valve at the bottom, which afterwards falling again, prevents the descent of the water that way. Then the first receiver being, at the same time, emptied of its air, push the handle of the regulator, and the steam which rises from the boiler will act upon the surface of the water contained in the first receiver, where the force or pressure on it still increasing its elasticity, till it exceeds the weight of a column of water in another receiving-pipe, then it will necessarily drive up through the passage into the force-pipe, and eventually discharge itself at the top of the machinery.

After the same manner, though alternately, is the first receiver filled and emptied of water, and by this means a regular stream kept continually running out of the top of a force-pipe, and so the water is raised very often from the bottom of a mine, to the place where it is meant to be discharged.

It should be added, that after the machine begins to work, and the water has risen into and filled the force-pipe, it fills also a little cistern, and by that means fills another pipe, called the condensing-pipe, which may be turned either way, over any of the receivers, when either is thoroughly heated by, the steam, to condense it within, thereby producing a vacuum, which absorbs the water out of the well into the receiver, on the principle of a syphon. Also a little above the cistern goes another pipe to convey the water from the force-pipe into the lesser boiler, for the purpose of replenishing the great boiler, when the water in it begins to be almost consumed. Whenever there is occasion for this, the cock is to be turned which communicates between the force-pipe and the lesser boiler, to close it effectually; at the same time having put a little fire beneath the small boiler, which will grow hot; its own steam, which has no vent to escape, pressing on its surface, will force the water up another pipe, through an aperture in the great boiler, and so long will it run, till the surface of the water gets so low as to be beneath the bottom of the pipe of communication—then the steam and water running together, will cause the valve (called a clack) to strike, which will intimate to the operator that it has discharged itself into the greater boiler, and carried in as much water as is then necessary; after which, by turning a cock, as much fresh water is let in as may be necessary; and then, by turning another cock, new fresh water is let out of a recipient into the less boiler as before; and thus the engine is supplied without fear of decay, or any delay in the operations; and proper attention in the workmen is only necessary to prevent disorder in a machine so expensive and complicated.

Also, to know when the great boiler wants replenishing, turn the gauge-cock; if water comes out, it does not need a supply; but if steam alone, then the want of water is certain. The like with the cock with which the lesser boiler is prepared for the same purpose, when the same state will be marked by like results. In working this engine, very little skill, and less labour is required: *Attention* is the chief requisite; it is only to be injured by want of due care, extreme stupidity, or wilful neglect.

The engine described above, does not differ essentially from that first designed by the inventor, Captain Savery;

the chief alteration which now occurs, is only in some few slight particulars. For example, the original engine had only one boiler, and there was no ready means for supplying it with water, to remedy the waste occasioned by evaporation of steam, without stopping the action of the engine, whenever the boiler was emptied to such a degree as to risk burning the vessel. After it was replenished the machine had to remain idle till the steam was raised, thus causing an immense loss of time; which is remedied by the application of a second boiler.

The description of the engine formerly mentioned is transcribed from Mr. Savery's publication, "The Miner's Friend," and which had a subsidiary boiler, with water of a boiling heat, always ready to supply the large boiler; and the power of steam raised in it is employed to force the water into the larger boiler, to replace the waste occasioned by evaporation from that boiler; by this means the transposition of the feeding water is not only speedily performed, but being itself of a boiling heat, it is instantly ready to produce steam for carrying on the work. There is also one more grand improvement in the modern machine: the first engine was worked by four separate cocks, which the operator was compelled to turn separately at every change of stroke; if he turned them wrong, he was not only liable to damage the engine, but he prevented its effect, and, at the same time, lost a part of the operation: whereas, in the improved engine, the communications are made by a double sliding valve, or, as it has since been termed, regulator; that is, a brass plate, shaped like a fan, and moving on a centre within the boiler, so as to slide horizontally in contact with the under surface of the cover of the boiler, to which it is accurately fitted by grinding, and thus, at pleasure, opens or shuts the orifices, or entries, to the steam pipes of the two receivers alternately. This regulator acts with less friction than a cock of equal bore, and, by the motion of a single handle backwards, at once opens the proper steam pipe from one receiver, and closes that which belongs to the other receiver. Captain Savery, in his publication before noticed, describes the uses to which this machine may be applied, besides those before described, viz.—1, to serve water for turning all sorts of mills; 2, for supplying palaces, noblemen and gentlemen's houses with water, and affording the means for extinguishing fires therein, by the water so raised; 3, the supplying cities and towns with water; 4, draining fens and marshes; 5, for ships; 6, for draining mines of water; and 7, for preventing damps in mines.

Dr. Desaguliers, we conceive, ungenerously attacked Captain Savery's reputation, by alleging that this was not an original invention, and that he was indebted for the first idea to the previously mentioned plan of the Marquis of Worcester. Dr. Rees, with a generous liberality worthy his great critical discrimination, scientific skill, and general erudition, has, we think, ably defended the captain's character, by proving his ideas to have originated with himself; we have only an opportunity to notice the most prominent features in this justification, where Dr. Rees thus expresses himself. "We know that the Marquis of Worcester gave no hint concerning the *contractibility or condensation of steam, upon which all the merit of the modern engine depends.* The Marquis of Worcester's engine was actuated wholly by the elastic power of steam, which he either found out, or proved by the bursting of cannon in part filled with water; and not the least hint that steam so expanded, is capable of being so far contracted in an instant, as to leave the space it occupied in a vessel, and occasion, in a great measure, a vacuum."

Subsequent to the Marquis of Worcester's, and Captain Savery's original ideas, and also, subsequent to the perfection the captain had brought his machine to, M. Amonton, a native of France, invented a machine which he called a fire-wheel; but it does not appear that it was ever brought to that perfection to be conducive to real utility, although it was certainly very ingenious.

Also, M. Papin, a native of Germany, made some pretensions to what he alleged was an invention of his own, only it happened to appear, unfortunately for his claim, that he was in London, and present at the time when Captain Savery exhibited the model of his steam-engine to the Royal Society. He made some unsuccessful experiments, by order of his patron, the Landgrave of Hesse, which sufficiently proved that, if he was the inventor, he did not understand the nature of his own machine.

Not long after Savery had invented his engine, Thomas Newcomen, an ironmonger, and John Calley, a glazier, began to direct their attention to the employment of steam as a mechanic power. Their first engine was constructed about the year 1711. This machine still acted on the principle of condensing the steam by means of cold water, and the pressure of the atmosphere on the piston. It was found of great value in pumping water from deep mines; but the mode of its construction, the great waste of fuel, the continued cooling and heating of the cylinder, and the limited capacities of the atmosphere in impelling the piston downward, all tended to circumscribe its utility.

The steam-engine was in this state, when it happily attracted the attention of Mr. Watt, to whom the merit and honour is due, of having first rendered this invention available as a mechanical agent. We cannot illustrate the improvements of this ingenious individual better than by giving a short biographical sketch of him to whom the world is so much indebted.

James Watt was born at Greenock, an extensive seaport in the west of Scotland, on the 19th of January, 1736. His father was a merchant, and also one of the magistrates of that town. He received the rudiments of his education in his native place; but his health being then extremely delicate, as it continued to be to the end of his life, his attendance at school was not always very regular. He amply made up, however, for what he lost in this way, by the diligence with which he pursued his studies at home, where, without any assistance, he succeeded, at a very early age, in making considerable proficiency in various branches of knowledge. Even at this time it is said his favourite study was mechanical science, to a love of which he was probably in some degree led by the example of his grandfather and his uncle, both of whom had been teachers of mathematics, and had left a considerable reputation for learning and ability in that department. Young Watt, however, was not indebted to any instruction of theirs for his own acquirements in science, the former having died two years before, and the latter one year after he was born. At the age of eighteen he was sent to London, to be apprenticed to a maker of mathematical instruments; but in little more than a year the state of his health forced him to return to Scotland; and he never received any further instruction in his profession. A year or two after this, however, a visit which he paid to some relations in Glasgow, suggested to him the plan of attempting to establish himself in that city, in the line for which he had been educated. In 1757, he accordingly removed thither, and was immediately appointed mathematical instrument maker to the College. In this situation he remained for some years, during which, notwithstanding almost constant ill health, he continued both to prosecute his profession, and to labour in the general cultivation of his mind, with extraordinary ardour and perseverance. Here also he enjoyed the intimacy and friendship of several distinguished persons, who were then members of the University, especially of the celebrated Dr. Black, the discoverer of the principle of latent heat, and Dr. Robison, so well known by his treatises on mechanical science, who was then a student, and about the same age as himself. Honourable, however as his present appointment was, and important as were many of the advantages to which it introduced him, he probably did not find it a very lucrative one; and therefore, in 1763, when about to marry, he removed from his apartments in the University, to a house in the city, and entered upon the profession of a general engineer.

For this his genius and scientific attainments most admirably qualified him. Accordingly he soon acquired a high reputation, and was extensively employed in making surveys and estimates for canals, harbours, bridges, and other public works. His advice and assistance were sought for in almost all the important improvements of this description, which were now undertaken or proposed in his native country. But another pursuit, in which he had been for some time privately engaged, was destined ere long to withdraw him from this line of exertion, and to occupy his whole mind with an object still more worthy of its extraordinary powers.

While yet residing in the College, his attention had been directed to the employment of steam as a mechanical agent, by some speculations of his friend Mr. Robison, with regard to the practicability of applying it to the movement of wheel-carriages; and he had also himself made some experiments with Papin's digester, with the view of ascertaining its expansive force. He had not prosecuted the inquiry, however, so far as to have arrived at any determinate result, when the winter of 1763–4, a small model of Newcomen's engine was sent him by the Professor of Natural Philosophy, to be repaired, and fitted for exhibition in the class. The examination of this model set Watt upon thinking anew, and with more interest than ever, on the powers of steam. Struck with the radical imperfections of the atmospheric engine, he began to turn in his mind the possibility of employing steam in mechanics, in some new manner which should enable it to work with much more powerful effect. This idea having got possession of him, he engaged in an extensive course of experiments, for the purpose of ascertaining as many facts as possible with regard to the properties of steam; and the pains he took in this investigation were rewarded with several valuable discoveries. The rapidity with which water evaporates he found, for instance, depended simply upon the quantity of heat which was made to enter it; and this again, on the extent of the surface exposed to the fire. He also ascertained the quantity of coals necessary for the evaporation of any given quantity of water, the heat at which water boils, under various pressures, and many other particulars of a similar kind, which had never before been accurately determined.

Thus prepared by a complete knowledge of the properties of the agent with which he had to work, he next took into consideration, with a view to their amendment, what he deemed the two great defects of Newcomen's engine. The first of these was the necessity arising from the method employed to concentrate the steam, of cooling the cylinder, before every stroke of the piston, by the water injected into it. On this account, a much more powerful application of heat than would otherwise have been requisite was demanded for the purpose of again heating that vessel when it was to be refilled with steam. In fact, Watt ascertained that there was thus occasioned, in the feeding of the machine, a waste of not less than three-fourths of the whole fuel employed. If the cylinder, instead of being thus cooled for every stroke of the piston, could be permanently hot, a fourth part of the heat which had hitherto been applied would be found sufficient to produce steam enough to fill it. How then was this desideratum to be obtained? Savery, the first who really constructed a working engine, and whose arrangements, as we have already remarked, all showed a very superior ingenuity, employed the method of throwing cold water over the outside of the vessel containing the steam—a perfectly manageable process, but at the same time a very wasteful one; inasmuch as every time it was repeated, it cooled not only the steam, but the vessel also, which, therefore, had again to be heated, by a large expenditure of fuel, before the steam could be produced. Newcomen's method of injecting the water into the cylinder was a considerable improvement on this; but it was still objectionable on the same ground, though not to the same degree; it still cooled not only the steam, on which it was desired to produce that effect, but also the cylinder itself, which, as the vessel in which more steam was to be immediately manufactured, it was so important to keep hot. It was also a very serious objection to this last mentioned plan, that the injected water, itself, from the heat of the place into which it was thrown, was very apt to be partly converted into steam; and the more cold water was used, the more considerable did this creation of new steam become. In fact, in the last of Newcomen's engines, the rarefaction of the vacuum was so greatly improved from this cause, that the resistance experienced by the piston in its descent was found to amount to about a fourth part of the whole atmospheric pressure by which it was carried down, or, in other words, the working power of the machine was thereby diminished one-fourth.

After reflecting for some time upon all this, it at last occurred to Watt to consider whether it might not be possible, instead of continuing to condense the steam in the cylinder, to contrive that method of drawing it off, to undergo that operation in some other vessel. This fortunate idea having presented itself to his mind, it was not long before his ingenuity suggested to him the means of realising it. In the course of one or two days, according to his own account, he had all the necessary apparatus arranged in his mind. The plan which he devised was, indeed, an extremely simple one, and on that account the more beautiful. He proposed to establish a communication by an open pipe, between the cylinder and another vessel, the consequence of which evidently would be, that when the steam was admitted into the former, it would flow into the other to fill it also. If, then, the portion in this latter vessel only should be subjected to a condensing process, by being brought into contact with cold water, or any other convenient means, what would follow? Why, a vacuum would be produced here—into that, as a vent, more steam would immediately rush from the cylinder—that likewise would be condensed—and so the process would go on till all the steam had left the cylinder, and a perfect vacuum had been effected in that vessel, without so much as a drop of cold water having touched or entered it. The separate vessel alone, or the condenser, as Watt called it, would be cooled by the water used to condense the steam—and that, instead of being an evil, manifestly tended to promote and quicken the condensation. When Watt reduced his views to the test of experiment, he found the result to answer his most sanguine expectations. The cylinder, although emptied of its steam for every stroke of the piston as before, was now constantly kept at the same temperature with the steam (or 212 deg. Fahrenheit); and the consequence was, that one-fourth of the fuel formerly required, sufficed to feed the engine. But besides this most important saving in the expense of maintaining the engine, its power was greatly increased by the most perfect vacuum produced in the new construction, in which the condensing water, being no longer admitted within the cylinder, could not, as before, create new steam there while displacing the old.

Such, then, was the remedy by which the genius of this great inventor effectually cured the first and most serious defect of the old apparatus. In carrying his ideas into execution, he encountered, as was to be expected, many difficulties, arising principally from the impossibility of realising theoretical perfection of structure with such materials as human art is obliged to work with; but his ingenuity and perseverance overcame every obstacle. One of the things which cost him the greatest trouble was, how to fit the piston so exactly to the cylinder, as, without affecting the freedom of its motion, to prevent the passage of the air between the two. In the old engine this end had been obtained by covering the piston with a small quantity of water, the dripping down of which into the space below, where it merely mixed with the stream introduced to effect the condensation, was of little or no consequence. But in the new construction, the superiority of which consisted in keeping this receptacle for the steam always both hot and dry, such an effusion of moisture, although in very small quantities, would have occasioned material inconvenience. The air alone, besides, which in the old engine followed the piston in its descent, acted with considerable effect in cooling the lower part of the cylinder. His attempts to overcome this difficulty, while they succeeded in that object, conducted Watt also to another improvement, which effected the complete removal of what we have called the second radical imperfection of Newcomen's engine, namely, its non-employment for a moving power, of the expansive force of steam. The effectual way it occurred to him of preventing any air from escaping into the part of the cylinder below the piston, would be to dispense with the use of that element above the piston, and to substitute there likewise the same contrivance as below, of alternate steam and a vacuum. This was, of course, to be accomplished by merely opening communications from the upper part of the cylinder to the boiler on the one hand, and the condenser on the other, and forming it at the same time into an air-tight chamber, by means of a cover, with only a hole in it to admit the rod or shank of the piston, which might, besides, without impeding its freedom of action, be padded with hemp, the more completely to exclude the air. It was so contrived accordingly, by a proper arrangement of the cocks and the machinery connected with them; that, while there was a vacuum in one end of the cylinder, there should be an admission of steam into the other; and the steam so admitted now served, not only by its susceptibility of sudden condensation to create the vacuum, but also, by its expansive force, to impel the piston.

These were the great improvements which Watt introduced in what may be called the principle of the steam-engine, or, in other words, in the manner of using and applying the steam. They constitute, therefore, the grounds of his claim to be regarded as the true author of the conquest that has been obtained by man over this powerful element. But original and comprehensive as were the views out of which these fundamental inventions arose, the exquisite and inexhaustible ingenuity which the engine, as finally perfected by him, displays in every part of its subordinate mechanism, is calculated to strike us perhaps with scarcely less admiration. It forms undoubtedly the best exemplification that has ever been afforded of the number and diversity of services which a piece of machinery may be made to render to itself, by means solely of the various application of its first moving power, when that has once been called into action. Of these contrivances, however, we can only notice one or two, by way of specimen. Perhaps the most singular is that called the *governor*. This consists of an upright spindle, which is kept constantly turning, by being connected with a certain part of the machinery, and from which two balls are suspended, in opposite directions, by rods, attached by joints, somewhat in the manner of the legs of a pair of tongs. As long as the motion of the engine is uniform, that of the spindle is so likewise, and the balls continue steadily revolving at the same distance from each other. But as soon as any alteration in the action of the piston takes place, the balls, if it has become more rapid, fly further apart under the influence of the increased centrifugal force which actuates them; or approach nearer to each other in the opposite circumstances. This alone would have served to indicate the state of matters to the eye; but Watt was not to be so satisfied. He connected the rods with a valve in the tube by which the steam is admitted to the cylinder from the boiler, in such a way, that as they retreat from each other, they gradually narrow the opening which is so guarded, or enlarge it as they tend to collapse; thus diminishing the supply of steam when the engine is going too fast, and when it is not going fast enough, enabling it to regain its proper speed by allowing it an increase of aliment.

Again the constant supply of a sufficiency of water to the boiler is secured by an equally simple provision, namely, by a *float* resting on the surface of the water which, as soon as it is carried down by the consumption of the water to a certain point opens a valve and admits more. And so on through all the different parts of the apparatus, the various wonders of which cannot be better summed up than in the forcible and graphic language of a recent writer:—"In the present perfect state of the engine it appears a thing almost endowed with intelligence. It regulates, with perfect accuracy and uniformity, the *number of its strokes* in a given time, *counting*, or *recording* them moreover, to tell how much work it has done, as a clock records the beats of its pendulum; it regulates the *quantity of steam* admitted to work; the *briskness of the fire*; the *supply of water* to the boiler; the *supply of coals* to the fire; it *opens and shuts its valves* with absolute precision as to time and manner; it *oils its joints*; it *takes out any air* which may accidentally enter into parts which should be vacuous; and when any thing goes wrong, which it cannot of itself rectify, it *warns its attendants* by ringing a bell; yet, with all these talents and qualities, and even when exerting the power of six hundred horses, it is obedient to the hand of a child; its aliment is coal, wood, charcoal, or other combustible—it consumes none when idle—it never tires, and wants no sleep; it is not subject to malady when originally well made, and only refuses to work when worn out with age; it is equally active in all climates, and will do work of any kind; it is a water-pumper, a miner, a sailor, a cotton-spinner, a weaver, a blacksmith, a miller, &c., &c.; and a small engine, in the character of a *steam pony*, may be seen dragging after it on a rail-road a hundred tons of merchandise, or a regiment of soldiers, with greater speed than that of the fleetest coaches. It is the king of machines, and a permanent realisation of the *Genii* of Eastern fable, whose supernatural powers were occasionally at the command of man."

In addition to those difficulties which his unrivalled mechanical ingenuity enabled him to surmount, Watt, notwithstanding the merit of his inventions, had to contend for some time with others of a different nature, in his attempts to reduce them to practice. He had no pecuniary resources of his own, and was at first without any friend willing to run the risk of the outlay necessary for an experiment on a sufficiently large scale. At last he applied to Dr. Roebuck, an ingenious and spirited speculator, who had just established the Carron iron-works, not far from Glasgow, and held also at the same time a lease of the extensive coal-works at Kinneal, the property of the Duke of Hamilton. Dr. Roebuck agreed to advance the requisite funds, on having two-thirds of the profits made over to him; and upon this Mr. Watt took out his first patent in the beginning of the year 1769. An engine with a cylinder of eighteen inches diameter was soon after erected at Kinneal; and although, as a first experiment, it was necessarily, in some respects, of defective construction, its working completely demonstrated the value of Watt's improvements. But Dr. Roebuck, whose undertakings were very numerous and various, in no long time after forming this connexion, found himself involved in such pecuniary difficulties, as to put it out of his power to make any further advances in prosecution of its object. On this Watt applied himself for some years almost entirely to the ordinary work of his profession as a civil engineer; but at last, about the year 1774, when all hopes of any farther assistance from Dr. Roebuck were at an end, he resolved to close with a proposal which had been made to him through his friend, Dr. Small, of Birmingham, that he should remove to that town, and enter into partnership with the eminent hardware manufacturer, Mr. Boulton, whose extensive establishments at Soho had already become famous over Europe, and procured for England an unrivalled reputation for the arts there carried on. Accordingly an arrangement having been made with Dr. Roebuck, by which his share of the patent was transferred to Mr. Boulton, the firm of Boulton and Watt commenced the business of making steam-engines, in the year 1775.

Mr. Watt now obtained from parliament an extension of his patent for twenty-five years, in consideration of the acknowledged national importance of his inventions. The first thing which he and his partner did was to erect an engine at Soho, which they invited all persons interested in such machines to inspect. They then proposed to erect similar machines wherever required, on the very liberal principle of receiving, as payment for each, only one-third of the saving in fuel which it should effect, as compared with one of the old construction.

But the draining of mines was only one of the many applications of the steam-power now at his command, which Watt contemplated, and in course of time accomplished. During the whole twenty-five years, indeed, over which his renewed patent extended, the perfecting of his invention was his chief occupation, and notwithstanding a delicate state of health, and the depressing affliction of severe headaches, to which he was extremely subject, he continued throughout this period to persevere with unwearied diligence in adding new improvements to the mechanism of the engine, and devising the means of applying it to new purposes of usefulness. He devoted, in particular, the exertions of many years, to the contriving of the best methods of making the action of the piston communicate a rotary motion in various circumstances, and between the years 1781 and 1785, he took out four different patents for inventions having this in his view.

It is gratifying to reflect, that even while he was yet alive, Watt received from the most illustrious contemporaries, the honours due to his genius. In 1785, he was elected a Fellow of the Royal Society; the degree of

Doctor of Laws was conferred upon him by the University of Glasgow, in 1806; and in 1808, he was elected a member of the French Institute. He died on the 25th of August, 1819, in the 84th year of his age.

The beneficial results arising from the ingenuity of Watt have been surprising. The steam-engine has already gone far to revolutionise the whole domain of human industry; and almost every year is adding to its power and its conquests. In our manufactures, our arts, our commerce, our social accommodations, it is constantly achieving what, little more than half a century ago, would have been accounted miraculous and impossible. "The trunk of an elephant," it has been finely and truly said, "that can pick up a pin, or rend an oak, is as nothing to it. It can engrave a seal, and crush masses of obdurate metal like wax before it—draw out, without breaking, a thread as fine as gossamer, and lift a ship of war, like a bauble, in the air. It can embroider muslins, and forge anchors; cut steel into ribbands, and impel loaded vessels against the fury of the winds and waves."

Another application of it is perhaps destined to be productive of still greater changes on the condition of society, than have resulted from many of its previous achievements,—we refer to railroads. The first great experiment was the Liverpool and Manchester Railway, which was opened, we believe, in 1831, and practically demonstrated, with what hitherto almost undreamt of rapidity travelling by land may be carried on through the aid of steam. Carriages, under the impetus communicated by this the most potent, and at the same time the most perfectly controllable of all our mechanical agencies, can be drawn forward at the flying speed of thirty and thirty-five miles an hour. When so much has been already done, it would be rash to conclude that even this is to be our ultimate limit of attainment. In navigation, the resistance of the water, which increases rapidly as the force opposed to it increases, very soon set bounds to the rate at which even the power of steam can impel a vessel forward. But on land, the thin medium of the air presents no such insurmountable obstacles to a force making its way through it; and a rapidity of movement may perhaps be eventually attained here, which is to us even as yet inconceivable. But even when the rate of land travelling already shown to be quite practicable shall have become universal, in what a new state of society shall we find ourselves! A nation will then, indeed, become a community; and all the benefits of the highest civilization will be diffused equally over the land, like the light of heaven. This invention, in short, when fully consummated, will confer upon man as much new power and enjoyment as if he were actually endowed with wings.

The commerce of the kingdom has also greatly benefited by the introduction of this valuable auxiliary, as will be seen from the following extract from the "Working Man's Companion:"—

"The establishment of steam-boats between England and Ireland has greatly contributed to the prosperity of both countries. How have steam boats done this? They have greatly increased the trade of both countries. On the examination of Mr. Williams, before a Committee of the House of Commons, he stated that 'before steam-boats were established, there was little trade in the smaller articles of farming production, such as poultry and eggs. The first trading steam-boat from Liverpool to Dublin, was set up in 1824; there are now (1832) forty such boats between England and Ireland. The sailing vessels were from one week to two or three weeks on the passage; the voyage from Liverpool to Dublin is now performed in fourteen hours. Reckoning ten mile, for an hour, Dublin and Liverpool are one hundred and forty miles apart; with the old vessels taking twelve days as the average time of the voyage, they were separated as completely as they would be by a distance of two thousand eight hundred and eighty miles. What is the consequence? Traders may now have, from any of the manufacturing towns in England, within two or three days, even the smallest quantity of any description of goods;' and thus 'one of the effects has been to give a productive employment of the capital of persons in secondary lines of business, that formerly could not have been brought into action.'" Mr. Williams adds, 'I am a daily witness to the intercourse by means of the small traders themselves between England and Ireland. Those persons find their way into the interior of England, and purchase manufactured goods themselves. They are, of course, enabled to sell them upon much better terms in Ireland; and I anticipate that this will shortly lead to the creation of shops and other establishments in the interior of Ireland for the sale of a great variety of articles which are not now to be had there.'

"And how do the small dealers in English manufactured goods find purchasers in the rude districts of Ireland for our cloths and our hardware? Because the little farmers have sent us their butter and eggs and poultry, and have either taken our manufactures in exchange, or have taken back our money to purchase our manufactures, which is the same thing. Many millions of eggs, collected amongst the very poorest classes, by the industry of the women and children, are annually sent from Dublin to Liverpool. Mr. Williams has known fifty tons, or eight hundred and eighty thousand eggs, shipped in one day, as well as ten tons of poultry; and he says this is quite a new creation of property. It is a creation of property that has a direct tendency to act upon the condition of the poorest classes in Ireland; for the produce is laid out in providing clothes for the females and children of the families who engage in rearing poultry and collecting eggs. Thus the English manufacturer is bettered, for he has a new market for his manufactures, which he exchanges for cheap provisions; and the dealer in eggs and poultry has a new impulse to this branch of industry, because it enables him to give clothes to his wife and children. This exchange of benefits—this advancement in the condition of both parties—this creation of produce and of profitable labour—this increase of the number of labourers—could not have taken place without machinery. That machinery is the carriage which conveys the produce to the river, and the steam-boat which makes a port in another country much nearer for practical purposes, than the market town of a thinly peopled district. A new machinery is added; the steam-carriage running on the railroad, as one of the witnesses truly says, 'is like carrying Liverpool forty miles into the interior, and thus extending the circle to which the supply will be applicable.' The last invention perfects all the inventions which have preceded it. The village and the city are brought close together in effort, and yet retain all the advantages of their local situation; the port and the manufactory are divided only by two hours distance in time, while their distance in space affords room for all the various occupations which contribute to the perfection of either. The whole territory of Great Britain and Ireland is more compact, more closely united, more accessible than was a single county two centuries ago."

The communication between England and Ireland has greatly increased since the above remarks were written, in 1832. There are now upwards of four hundred steam-boats sailing between Ireland and Great Britain, and of late years the largest export from that unfortunate country consists of her starving population, who, true enough, find their way into the interior of England, but not with the intention of purchasing manufactured goods, but of being employed in the manufacturing of them. We believe that our mechanical readers, at least, will agree with us, when we say that the benefit has not been reciprocal. England, for her share, has been burthened with a pauper population, and her sons deprived of their employment, by the immense immigration that has of late years taken place. Poor rates are multiplied to an extent hitherto unheard of, and our streets swarming with beggars—and those of the most importunate class. So much was this the case, that in 1847 and 1848, Liverpool was inundated with paupers from the sister country to such a degree, that her authorities were compelled to petition government to put an end to the nuisance, and to grant them assistance to prevent the death of so many thousands of their fellow-men from dying for want; the poor-rates were so increased that the ratepayers with justice complained. And we question much if ever the English manufactures have been so much benefited by the commerce as the foregoing quotation would lead us to believe. That we have been supplied with enormous quantities of provisions we cannot deny; but that the payment of these was taken back in our cloths and our hardware is very questionable. That the money was taken back there can be little doubt, not for the purpose, however, of buying clothes for the wives and children of those families whose industry had supplied us with eggs and poultry, but for supplying the insatiate wants of their profligate landlords, who were squandering the subsistence of the needy peasantry in another land. If any class of men have obtained benefit by means of this increased and speedy communication between the two countries, it assuredly is the absentee Irish landlord.

MILLS.

CORN Mills are of very ancient origin, and it may not be uninteresting to our readers to learn something of the customs of our forefathers with regard to them; to which we will subjoin such modern improvements as the more advanced state of the arts have enabled the moderns to achieve, and to excel the imperfect information of the ancients in mechanical sciences.

In support of the antiquity of grinding corn, we may go as for back as the days of the patriarch Abraham, who, we are informed in Genesis xviii. 6, "hastened into the tent unto Sarah, and said, Make ready quickly three measures of fine meal, knead it, and make cakes upon the hearth." To this we may add, that it appears in a subsequent text, Numbers xi. 8, that manna was ground like corn. The earliest instrument for this purpose seems to have been the mortar, which was retained long after the introduction of mills, properly so called: because they were most probably at first very imperfect. In process of time the mortar was made ridged, and the pestle notched at the bottom, by which means the grain was rather grated than pounded.

A passage in Pliny, which has not as yet had a satisfactory interpretation, renders this conjecture probable. In time a handle was added to the top of the pestle, that it might be more easily driven round in a circle, whence this machine at first was called *mortarium*, by this means assuming the name of a hand-mill. Such a mill was so called from rubbing backwards and forwards; and varied but little from those used by our colour-grinders, apothecaries, potters, and other artisans. From expressions in the sacred volume, we may rationally infer that it was customary to have a mill of this sort in every family.

Moses having forbidden to take such instruments for a pledge; for that, says he, "No man shall take the nether or the upper millstone to pledge: for he taketh a man's life." It is observed by Michaelis, on this passage, that a man could not then grind, consequently could not bake the necessary daily bread for the family.

Grinding was then the employment of the women, particularly of female slaves, as at present in those countries which are uncivilised: the portion of strength required for the operation, therefore, could not have been great; but afterwards the mills were driven by bondsmen, whose necks were placed in a circular machine of wood, so that they could not put their hands to their mouths or eat of the meal. This must have been an interesting link between the hand and the horse-mill.

In course of time shafts were added to the mill, that it might be driven by cattle, which were then blindfolded. The first cattle mills were called *molae jumentaria*, which had, probably, only a heavy pestle like the hand-mill; but it is conjectured, that it must have been soon remarked, that the labour would be more easily accomplished, if, instead of the pestle a large heavy cylinder was employed. A competent judge has, however, believed that the first cattle mills had not a spout or trough as ours have; at least those hand-mills Tournefort saw at Nicaria, consisted only of two stones; but the meal issued through an opening in the upper one, and fell upon a board or table, on which the lower one rested.

The upper millstone they called *meta*, or *turbo*; and the lower one *catillus*: the name of the first also signified a cone with a blunt apex, whence it has been thought by some, that corn was first rubbed into meal, by rolling one stone upon another, as painters now grind colours with a muller. This is not improbable, as present practice among barbarous people fully proves. It is also apparent that the upper millstone was substituted for the pestle, which action may have lent it a name, when they called it *meta*.

Professor Beckmann has followed Gori in his description of an antique gem, engraved on red jasper, upon which appears "the naked figure of a man, who in his left hand holds a sheaf of corn, and in the right a machine that in all probability is a hand-mill. Gori considers the figure as a representation of the god Eunostus, who was the god of mills. The machine which Eunostus seems to exhibit, or to be surveying himself, is, as far as one can distinguish, (for the stone is scarcely half an inch in size), shaped like a chest, narrow at the top, and wide at the bottom. It stands upon a table, and in the bottom there is a perpendicular pipe, from which the meal, also represented by the artist, appears to be issuing. Above, the chest or body of the mill has either a top with an aperture, or perhaps a basket sunk into it, from which the corn falls into the mill. On one side, nearly about the middle of it, there projects a broken shank, which, without overstraining the imagination, may be considered as a handle, or that part of the mill which some call *mobile*. Though this figure is small, and though it gives very little idea of the internal construction, one may, however, conclude from it that the roller, whether it was of wood or of iron, smooth or notched, did not stand perpendicularly, like those of our coffee mills, but lay horizontally, which gives us reason to conjecture a construction more ingenious than that of the first invention. The axis of the handle had, perhaps, within the body of the mill, a crown wheel, that turned a spindle, to the lower end of the perpendicular axis of which the roller was fixed. Should this be admitted, it must be allowed also, that the hand-mills of the ancients had not so much a resemblance to the before-mentioned colour mills as to the philosophical mills of our chemists; and Langelott, consequently, will not be the real inventor of the latter. On the other side, opposite to where the handle is, there arise from the mill of Eunostus two shafts, which Gori considers as those of a besom and shovel, two instruments used in grinding; but as the interior part cannot be seen, it appears to me doubtful whether these may not be parts of the mill itself."

In the commencement of the last century, the remains of a pair of Roman millstones were found at Adel, in Yorkshire. One of these stones, twenty inches in breadth, is thicker in the middle than at the edge, consequently one side is convex; the other was of the same size, but as thick at the sides as the other was in the centre; the traces of notching were discoverable.

Enough, may, perhaps, have been said concerning this original invention; therefore this article will not be encumbered with quotations of all those passages relative to mills, which are found in ancient authors, as they would afford but little additional information. Neither will mythological records be disturbed to inquire to which deity or hero the invention was originally attributed; or to ascertain the descent of Milantes, whom Stephanus distinguishes by that honour, or how those millstones were constructed which are alleged to have been built by Myletes, son of Lelex, King of Laconia; but we shall proceed to the invention of Water-Mills.

These appear to have been introduced about the period of Mithridates, contemporary with Cæsar and Cicero. Strabo, relating that there was a water-mill near the residence of the Pontian king, that honour has been ascribed to him; but so far is this remote from certainty, that nothing can be inferred from thence, other than that water-mills at that period were known in Asia. Pomponius Sabinus informs us, that the first water-mill seen at Rome was erected on the banks of the Tiber, a little before the time of Augustus; but of this there is no other proof than his simple assertion: he having taken the greater part of his remarks from the illustrations of Servius, he must have had a more perfect copy of that author than any now remaining, and from these his information might have come.

The most certain proof we have that Rome had water-mills in the time of Augustus, is, that Vitruvius has told us so; but those mills were not corn-mills, they were hydraulic engines, which he describes in his works. From whence we learn that the ancients had wheels for raising water, which were driven by being trod upon by men; the usual employment for criminals, as may be learnt from Artemidorus. Also from a pretty epigram of Antipater; "Cease your work, ye maids, ye who laboured in the mill; sleep now, and let the birds sing to the ruddy morning; for Ceres has commanded the water nymphs to perform your task; these, obedient to her call, throw themselves on the wheel, force round the axle-tree, and by these means the heavy mill." Antipater lived at the period of Cicero. Palladius, also, with equal clearness, speaks of water-mills, which he advises to be built on estates where is running water, in order to grind corn without men or cattle.

It likewise appears that the water-wheels to which Heliogabalus directed some of his friends and parasites to be tied, cannot be considered to be mills for the purpose of grinding corn; for these, as well as the *haustra* of Lucretius were probably like those machines for raising water, which are spoken of by Vitruvius as *hydraulic*.

It is, however, on the authority of Pompinius Sabinus, before-cited, that both wind and water mills were known to have been in Italy, and even the latter in Rome, in the days of Augustus. However, about twenty-three years after the death of Augustus, when Caligula seized every horse from the mills, to convey effects he had in contemplation to take from Rome, the public were much distressed for bread; whence we must infer that water-mills must have been very rare. Even three hundred years after Augustus, cattle mills were so common in that city, that their number amounted to three hundred; mention of them, and of the hand-mills, often occurs for a long time after. It is not their use we inquire after, it is enough for us to know that they existed.

We now come to another period, when we are informed that *public mills* were first introduced, which occurs in the year 398, mention being made of them in that year, which also clearly shows that they were then newly-established; which establishment was found necessary to be protected by laws made in their favour. The orders for that purpose were renewed more than once, and made more secure by Zeno, towards the end of the fifth century. It may be properly remarked, that in the whole code of Justinian, the least mention of wooden pales or posts is not made, which occurs in all the new laws,—and which, it appears, when there were several mills on the same stream, occasioned so many disputes then, as well as in after times. The mills at Rome were erected on those canals which conveyed water to the city; and because these were employed in several arts, and for many purposes, it was ordered that, by dividing the water, the mills should always be kept going; but as they were driven by so small a quantity of water, they probably executed very little work; and for this reason, but probably on account of the great number of slaves, and the cheap rate at which they were maintained, these noble machines were not so much used, nor were so soon brought to perfection, as under other circumstances they might have been. It appears, however, that after the abolition of slavery, they were much improved, and more employed, and to this a particular incident seems, in some degree, to have contributed.

When Vitiges, King of the Goths, besieged Belisarius in Rome, in the year 536, and caused the fourteen large expensive aqueducts to be stopped, the city was reduced to great distress; not from want of water, in general, because it was secured against that inconvenience by the Tiber; but by the loss of that water which the baths required, and, above all, of that necessary to drive the mills, which were all situated on these canals. Horses and cattle, which might have been employed upon grinding, were not to be found; but Belisarius fell upon the ingenious contrivance of placing boats upon the Tiber, on which he erected mills that were driven by the current. This experiment was attended with complete success; and as many mills of this kind as were necessary were constructed. To destroy these, the besiegers threw into the stream logs of wood, dead bodies, &c., which floated down the river into the city; but the besieged, by making use of booms to stop them, were enabled to drag them out before they could do any mischief. This seems to have been the origin of *floating-mills*, no record of them appearing previously. By these means the use of water-mills became very much extended; for floating-mills can be constructed almost upon any stream, without forming an artificial fall; they may be stationed at the most convenient places, and they rise and fall of themselves with the water.—They are, however, attended with these inconveniences: they require to be strongly secured; they often block up the stream too much, and move slowly; and they often stop when the water is too high, or when it is frozen.

After this improvement, the use of water-mills was never laid aside or forgotten, but was soon made known all over Europe; and passages innumerable might be quoted, in every century, to prove their continued use. The Roman, Salic, and other laws, constantly provided for the security of these mills, and defined a punishment for such as destroyed the sluices, or stole the mill-irons. It is said, however, that there were water-mills in France and Germany a hundred years before these laws had existence.

At Venice, and other places, there were erected mills which regulated themselves by the motion of the waters, and which were regulated by the flowing and ebbing of the tide, and which every six hours changed the motion of the wheels. Of this species of mills, a new invention, or, perhaps, rather an improved one, was made in London, called a tide-mill, an engraving of which may be seen in "The Advancement of Arts, Manufactures, and Commerce," London, 1772.

Zanetti is said to have shown, by some old charters, that such mills existed about the year 1044; but with still more certainty in 1078, 1079, and 1107.

It appears, however, that hand and cattle mills were in most places retained, after the use of proper watermills, particularly in convents. They were used, because the otherwise lazy monks found the exercise they afforded beneficial to their health. Likewise the legends of popish mythology are full of the miracles which have been wrought at these mills.

A modern author of credit impeaches the veracity of Pomponius Sabinus after he had previously quoted his authority, and likewise after he had said that he bore a good character, in a popular work, by charging him with improbability, nay, positive falsehood, and alleging that the Romans had no wind-mills. It should be noticed, without venturing to decide upon the point, that he has adduced no authority for such allegation, and that he only concludes so, by inference, as upon the authority of Vitruvius; that mechanist, he says, in enumerating all moving forces, does not mention wind-mills. But, for the sake of candour, was not the one as liable to err as the other? He also says, that neither Seneca nor St. Chrysostom mention wind-mills; and is unmercifully severe upon an old Bohemian annalist who speaks of wind-mills so early as 718. But he is all along bringing his forces to prove, that wind-mills had first existence in his own district, Germany; that they were then invented; and, perhaps, because he is of that country. It is somewhat remarkable that scarcely any invention of any consequence has occurred since that of printing, but the honour has been claimed by the natives of Germany.

Mabillon mentions a diploma of the year 1105, in which a convent in France is allowed to erect water and wind-mills, *molendina ad ventum*.

Bartolomeo Verde proposed to the Venetians in 1332, to build a wind-mill. When his plan had been examined, he had a piece of ground assigned him, which he was to retain if his undertaking succeeded within a specified time. In 1373, the city of Spires caused a wind-mill to be erected, and sent to the Netherlands for a person acquainted with the method of grinding by it. A wind-mill was also constructed at Frankfort, in 1442; but it does not appear to have been ascertained whether there were any there before.

About the twelfth century, in the pontificate of Gregory, when both wind and water-mills became more general, a dispute arose whether mills were titheable or not. The dispute existed for some time between the persons possessed of mills and the clergy; when neither would yield. At length, upon the matter being referred to the pope and sacred college, the question was, (as might have been expected when interested persons were made the arbitrators,) determined in favour of the claims of the church.

There was one inconvenience attending wind-mills, which might be obviated in other mills: the mill was useless unless the wind was in a particular direction. To remedy this, various modes were tried; at first, the mill was fixed on a floating body in the water, which might be turned to any wind. The next improvement consisted in turning the body of the mill to meet the direction of the wind; this was effected by two modes: first, the whole building is constructed in such a manner as to turn on a pivot below; this method is said to have been invented in Germany, and is called the German mode: second, the building is formed so as to turn on the roof, with the shafts supporting the sails only; this is called the Dutch mode, being invented by a Fleming about the middle of the sixteenth century. This is the mode principally adopted in England.

Although in the earliest ages of the world men might have been, perhaps, satisfied with having their corn reduced to a mealable form alone; yet after this had been with care effected, then they thought of improving upon this conveniency, and separating the farinaceous part from the bran and husks. This was certainly desirable; therefore they bolted it in a sieve with a long handle attached to it, with a hair, or fine lawn lining; this was common in this country till within the last sixty or eighty years; but by degrees, opportunities of improvement in the mechanism of mills suggested to some mechanic the idea of constructing what is now called bolting mills, applied to the mill for grinding, and wrought at the same time by appropriate machinery.

It appears that sieves of horse-hair were first used by the Gauls, then those of linen by the Spaniards. The mode of applying a sieve in the form of an extending bag to catch the meal as it fell from the stones, and of causing it to be turned and shaken, was first made known in the beginning of the sixteenth century.

The best bolting cloths are universally allowed to be manufactured in England; they are made of wool of the longest and the best kind, peculiarly prepared; being first well washed and spun to a fine and equal thread; which, before it be scoured, must be scalded in hot water to prevent its shrinking. The web must be then stiffened; it is in this we possess an advantage which others cannot attain. Our bolting cloth is stiffer, as well as much smoother, than any foreign manufacture. So jealous are our German neighbours of this, that they have established manufactories in several places at a great expense, and under very peculiar regulations, for its fabrication. After all, they are compelled to confess, that theirs will not wear above three weeks in a flour manufactory, whereas ours will continue well three months in equal exposure to friction and ordinary wear.

For some years past, the French have been extolled for a mode of grinding, called *mouture economique*; that were we not aware such had been practised in ancient Rome, it might be conceived to form an important epoch in the miller's art. This process, however, is not new; it consists in first grinding the wheat not so fine as might be required for ordinary purposes; afterwards putting the meal several times through the mill, and sifting it with various sieves. It should seem this method was practised in ancient Rome; for Pliny, who took care to inform himself of most things, tells us, that in his time they had, at least, five different kinds of flour, all procured from the same corn. It appears, that the ancient Romans had advanced very far in this art, as well as in that of baking, &c., from what may be collected from its economical polity preserved by Pliny and others. Whence it may be fairly inferred, they knew how to prepare from corn more kinds of meal, and from meal more kinds of bread, than the moderns even now are acquainted with.

Pliny reckons that bread should be one-third heavier than the meal used for baking it: this proportion it appears, was known in Germany nearly a century and a half ago, and discovered from experiments on bread made at different times. German bakers, although they may have been occasionally mistaken, have always undoubtedly given more bread than meal. It appears that in latter periods, the art of grinding, as well as baking, has declined very much in Italy; and their bread, although produced from the finest grain in the world, is altogether bad when manufactured by Italians. On this account, bakers from Germany it seems, are generally employed in public baking-houses, as well at Rome as in Venice. Bakers of that people are generally settled at those places, where they have been in the habit of manufacturing that article for the principal inhabitants, for upwards of three hundred years.

From Beckmann's History, it would appear that the *mouture economique* of the French has been known to the Germans for more than two hundred years. Many were the attempts, repeatedly enforced, to deter the experiments made, from time to time, by the French experimentalists, to perfect this article previous to its being accomplished. In this, the French suffered themselves to be taught by prejudice and directed by ignorance. Numerous and judicious were the experiments made by the scientific and philosophic of that people to produce the most in quantity and best in quality from a definite quantity of grain, at which the ignorant of their species suffered their prejudice to revolt, and the powerful readily come into the mode of thinking of the vulgar, to whom they lent their aid, to effect what Heaven in revelation had commanded, viz: "Give not that which is holy unto dogs, neither cast ye your pearls before swine, lest they trample them under their feet, and turn again and rend you." Mat. vii. 6.

It will, from the succeeding statement, that in using the language which has just appeared, circumstances sanctioned us. The clergy of the chapel royal, and parish church at Versailles, sent their wheat in the beginning of last century to be ground at an adjacent mill: according to custom, it was put through the mill only once, and the bran, which yet contained much flour, was sold for fattening cattle. This miller having, however, in process of time learnt the process of the *mouture economique*, purchased the bran from these ecclesiastics, and found that it yielded him as good flour as they had procured from the whole wheat. The miller, at length, is presumed, in a qualm of conscience, to have regretted cheating those holy men; he accordingly discovered to them the secret, and gave them afterwards fourteen bushels of flour from their wheat, instead of eight, which he had only furnished them before. This voluntary discovery of the miller was made in 1760; and it is probable the same discovery was made at the same time by others.

A baker, named Malisset, proposed to the lieutenant-general of the French police to teach a method by which people could grind their corn with more advantage; and experiments were accordingly made and succeeded. A mealman of Senlis, named Buquet, having the inspection of the mill belonging to the large hospital at Paris, made the same proposal: the result of his experiments, made under the direction of the magistrates, was printed. The investigation of this art was now taken up by men of learning and science, who gave it a suitable denomination;

explained it, made experiments and calculations upon it, and at the same time recommended it so much, that the *mouture economique* engaged the attention of all magistrates throughout France. Its government sent Buquet to Lyons in 1764, to Bourdeaux in 1766, to Dijon in 1767, and to Mondidier in 1768. The benefit which France derived from that trouble, shows that it was not taken in vain. Previous to that period, a Paris *setier* yielded from eighty to ninety pounds of meal, and from one hundred and fifty to one hundred and sixty pounds of bran; but the same quantity now yields one hundred and eighty-five pounds, and according to the latest improvements, one hundred and ninety-five pounds of meal. In the time of St. Louis, from four to five *setiers* were reckoned necessary for the annual maintenance of a man; these were scarcely sufficient; as many were allowed to the patients in hospitals; and such were the calculations made in the sixteenth century. When the miller's art was everywhere improved, the four *setiers* were reduced to three and a half, and from the latest improvements, they do not exceed two.

From mills which only force the farinaceous parts from the husk, thereby rounding the grain, the common denomination of *barley mills* comes, from such mills being used in the manufacture of pearl barley. In their construction, these mills differ but little from wheat-mills, and the machinery for the former is generally added to the latter. The grand specific distinction is, that the millstone is rough hewn round its circumference, and in the stead of a lower stone, there is generally a wooden case; the middle lined with a plate of iron, pierced like a grater with holes, the sharp edge of which turns upwards. The barley is thrown upon the stone, which, as it turns round, frees it from the husk, and rounds it; after which, it is put into sieves and sifted.

So long as the policy of governments was blind to the interests of men, and so long as the griping avarice of a few was permitted to lay the free-born of their species under the most severe contributions, so long were permitted to build mills only, who had obtained a regal license for that purpose. But, thank heaven! that ray of light it has lent generally to man, has, in some sort, illuminated even the minds of ministers and their tyrannical masters, to curtail that spirit which had cast the fetters of vassalage given by feudal tyranny to its upstart dependants. Men were left, at length, to improve their property according to their pleasure: since which period, more mills have been erected for the convenience of the species. This privilege, it appears, was not prohibited by the Roman laws; those irradiations of superior intellect well appreciated human rights. It was not till the darkness of the middle ages had obscured the mental hemisphere, that any person was presumed to possess a superiority over others, and to abridge the small portion of general happiness that the favoured of fortune might add to his satiety. During those days of universal darkness, numberless were the evils which men suffered, and among them the present object of our consideration was not the least; frequently having to travel for miles to a mill to procure the necessary manufacture of so essential an article to human life as bread.

Let us not be decoyed, however, by the resentment produced by the spirit of human oppression, beyond the bounds prescribed by reason, to inveigh against such ordinance when public and general utility ever was consulted; and certain public streams were by wise laws to be kept free from individual encroachments with impunity. It is not against the dictates of sober reason we declare hostility, but the gross abuse of power.

A time there was, when human baseness in princes and potentates, their vassals doubtless aping the manners of their masters, claimed as their right not only the common element of water, but also that of air! A curious incident related by Jargow, and detailed by Professor Beckmann, as follows, establishes the insolence of upstart men:—"In the end of the fourteenth century, the monks of the celebrated but long since destroyed monastery of Augustines, at Windshiem, in the province of Overyssel, were desirous of erecting a wind-mill not far from Zwoll; but a neighouring lord endeavoured to prevent them, declaring that the wind in that quarter belonged to him. The monks, unwilling to give up their point, had recourse to the Bishop of Utrecht, under whose jurisdiction the province had continued since the tenth century. The bishop, highly incensed against the pretender, who wished to usurp his authority, affirmed, that the wind of the whole province belonged only to him; and, in 1391, gave the convent express permission to build a wind-mill wherever they thought proper."

Without the convenience of human ingenuity heaven had sent the blessing of life in vain; we have, under this impression, therefore, bestowed much time on this article, from a conviction of its vital importance to the necessities of human existence.

SAW-MILLS.

THE invention of the plumb-line and saw, with other useful articles in mechanics, and handicrafts, are usually ascribed to that great—that universal genius—Dædalus: although others give the merit to one Talus, the nephew of Dædalus, and say, that the discovery was made under the following circumstances:—Talus, they tell us, having found the jawbone of a snake, cut a piece of wood in two with the teeth; thence, they say, he invented the saw; his maternal uncle and master, they add, was so jealous of this invention, that he murdered the young man; and the mode of the discovery of the murder is accounted for in this manner:—some persons saw Dædalus covering up the grave of his victim, and asked what he was doing? "Oh," says he, "I am only burying a snake." How much credit may be due to this relation, we do not take upon ourselves to determine. Pliny, as well as Seneca, were of the former opinion; whilst Diodorus Siculus, and others, hold the latter. The youth is named by some Perdix. However, it appears to rest between these two, no other claimant appearing. Ovid says, it was not the jaw of a snake, but the back-bone of a fish. The former, however, appears to be the most rational opinion as to its origin, as it is conjectured that the vertebræ would not be sufficiently strong, and the joints are too far apart, as well as too large.

The Grecian saw is said to have been much the same as that instrument which the moderns now use. This idea is corroborated by an ancient painting discovered in Herculaneum; likewise from an antique representation of this instrument, given by the celebrated Montfaucon.

The preceding observations, however, have relation to the subject of this article only, inasmuch as they are introductory to what follows.

The most beneficial and ingenious improvement that has been made in saws was the invention and introduction of machinery, called saw-mills, which, in woody countries, as well as for delicate and fine veneers, are of the greatest utility; in the former case, wood forms the chief article of commerce where labourers are scarce; in the latter, it may be cut nearly as thin as a sheet of paper. These saw-mills also finish flooring deals, grooved, dovetailed, and planed on both sides, at the rate of two deals, of twenty feet each, in a minute! They are commonly worked in this country by means of steam-engines; in woody countries they are generally erected on the banks of rivers, the water of which propels the machinery.

It is said they were invented in Germany, as far back as the fourth century, upon the smaller river Roer; for, although Ansonius speaks of water-mills, for cutting stone, he says nothing of mills to cut timber. The art of cutting marble with a saw is very ancient; Pliny thinks it was invented in Caria; at least, he knew of no place or building, incrusted with marble, older than the palace of King Mausolus, at Helicarnassus. Vitruvius also names the circumstances, although he uses different terms for expressions of the same sense. He commends the beauty of its marble, whilst Pliny speaks of its different kinds: the former viewed it as an architect, whilst the latter inspected it as a naturalist. It also does appear, from other writers, that the harder and precious kinds of stones were cut in the same manner; as Pliny speaks of a building adorned with agate, cornelian, lapis-lazuli, and amethysts. Yet there is no mention made of mills for cutting wood; or, admitting they had been invented, it is probable they shared the fate of many other useful inventions,—had been forgotten, or else some considerable modern improvement had been made in their construction.

Since the period of the first invention, they have been erected in various parts of Europe and America. There appears to have been one erected in the vicinity of Augsburg, as early as 1337; at Erlinger, in 1417.

Upon the discovery of the island of Madeira, in 1420, the Infanta Henry sent settlers there, and caused European fruits of every kind to be carried there; and amongst other productions, saw-mills and other machinery to cut the valuable timber found there into portable pieces, which were afterwards transported to Portugal. In 1724, the city of Breslau had a saw-mill which produced the yearly rent of three marks. In 1490, the magistrates of Erfurt purchased a forest, and built a mill of this description. In Norway, a country covered with wood, there was one built in 1530. This mode of manufacture was called the new art; and because the exportation of deals was by that means increased, a royal impost was introduced by Christian III. in 1545, called the deal-tythe. Soon after Henry Ranzau caused the first mill to be erected at Holstein. In the year 1555, the Bishop of Ely, being ambassador from the Princess Mary of England to the court of Rome, saw a saw-mill in the neighbourhood of Lyons: the writer of his travels thought it worthy of particular description:—"The saw-mill is driven by an upright wheel; and the water that makes it go is gathered whole into a narrow trough, which delivereth the same water to the wheels. This wheel hath a piece of timber put to the axle-tree end, like the handle of a brooch, and fastened to the end of the saw, which being turned with the force of the water, hoisteth up and down the saw, that it continually eateth in, and the handle of the same is kept in a rigall of wood from swerving. Also the timber lieth as it were upon a ladder, which is brought by little and little to the saw with another vice." In the sixteenth century, there was a grand improvement made in this machine by having several saws affixed to one beam, by which timber could be cut into several planks or boards, and of any thickness, at the same time. There was one of these at Ratisbon, upon the Danube, in 1575.

In England saw-mills were at first received with as little encouragement as printing met with in Turkey, and from the same motive. When the attempt was made to introduce them it was said the sawyers would be deprived of bread. For this reason it was found necessary to abandon a saw-mill erected by a Dutchman, near London, in 1663. However, in the year 1700, a gentleman of the name of Houghton laid before the nation the advantages to be derived from them; but he expressed his apprehension that it might cause a commotion among the people. What he feared, actually came to pass; for, on the erection of one by a wealthy timber merchant, by the desire of the society for the promotion of arts, in 1767, to be propelled by the wind, under the direction of James Stansfield, who had learnt the method of constructing them in Holland and Norway, a foolish mob assembled and pulled it to pieces. Many years previous to this there had been a similar mill erected in Scotland. There is now hardly a town of any importance in the kingdom but what has one or more saw-mills in operation.

FORKS.

THE fork is an article of every-day use amongst us, and on that account little thought of; still the short space we intend to occupy with this subject may, perhaps, convey a little information to many of our readers unknown to them before, or, at least, unthought of.

There is not the least room to suppose the ancients were at all acquainted with this little table utensil, now so necessary to our own comfort and convenience, to say nothing of our ideas of cleanliness. Pliny, who enumerated most things natural, physical, philosophical, and economical, makes no mention of them; nor does it occur in any other writer of antiquity; neither does Pollux speak of it in the very full catalogue which he has given of things necessary for a table.

Neither the Greeks or Romans had any name in the least applicable to its use, either direct or by inference, where it can be asserted that such an instrument was intended. The ancients had, it is true, in Greece, their *creagra*. In Rome, their *furca, fuscina, furcilla*, &c.: the Grecian instrument somewhat resembled a rake of an ordinary construction, and calculated for the purpose of taking meat out of a boiling pot, constructed in the shape of a hook, or rather the bent fingers of the hand.

With reference to the Roman names, the first two were undoubtedly applied to instruments which approached nearer to our furnace and hay forks.—The trident of Neptune is also called *fuscina*. The furcilla was large enough to be employed as a weapon of defence. The present Latin name for a fork, *fusinula*, is not to be found in any of the old Latin writers.

It is the opinion, we understand, of a learned Italian writer, that the ancient Romans used the instruments they called *ligulæ*, instead of forks. Now those instruments had some distant resemblance to our teaspoons. Hence we must conclude that they and our ancestors used no forks, because, had they had anything answering the purpose, even in effect, it must undoubtedly have had a name.

In the East, we understand it was, and still is, customary to dress their victuals until they become so tender as to be easily pulled in pieces. We are told by modern travellers, that if an animal be dressed before it has lost its natural warmth, it becomes tender and very savoury. This is the Oriental custom, and has been so from the most remote antiquity.

Fortunately, all articles of food were cut up in small pieces before they were served up at table; the necessity for which practice will appear, when we remember they usually took their meals in a recumbent posture upon beds. Originally, persons of rank kept an officer for the purpose of cutting the meat, who used a knife, the only one placed at table, which, in opulent families, had an ivory handle, and was ornamented with silver.

The bread was never cut at table; it needed it not, being usually baked thin, somewhat resembling the Passover cake of the Jews; this is not understood, however, to have been universal.

The Chinese use no forks; however, to supply them, they have small sticks of ivory, often of very fine workmanship, inlaid with silver and gold, which each guest employs to pick up the bits of meat, it being previously cut small. The invention of forks was not known till about two centuries ago in Europe, where people eat the same as they do now in Turkey.

In the New Testament we read of putting hands into the dish. Homer, as well as Ovid, mention the same custom.

In the quotation from the sacred writings, we observe that the guests had, it is presumed, no instrument to help themselves out of the common dish which contained the repast; for, upon the question being put of who was to betray the Saviour, the answer was given in the following quotation, "It is one of the twelve that dippeth with me in the dish."

In the passage cited from Homer, the phrase, according to the Latin translation, implies the same sense. And had the Romans been apprised of the utility of this instrument, or in fact of any substitute, there could have been no occasion for the master of the amorous art to have given his instructions to his pupils in nearly similar terms which we now use to children.

Although Count Caylus and Grignon both assert that ancient forks have been found, we still want further testimony. The former says, one with two prongs was found among some rubbish in the Appian Way, which he alleges to be of beautiful workmanship, terminating in the handle with a carved stag's foot. Notwithstanding the high reputation of that author, this assertion is not credited. The latter says, he found some in the ruins of a Roman town in Champagne; but he does not describe them, otherwise than to observe that one was of copper or brass, and the others of iron: and speaking of the latter, says, they appear to be table-forks, but are very coarsely made.

The truth seems to be that table-forks were first used in Italy, as appears from the book of Galeotus Martius, an Italian in the service of Matthias Corvinus, King of Hungary, who reigned from 1458 to 1490. Martius relates that at that period forks were not used at table in Hungary as in Italy; but that at meals each person laid hold of the meat with his fingers, and on that account they were much stained with saffron, usually put into sauces and soups. He praises the king for eating without a fork, conversing at the same time, and never dirtying his clothes.

In France, at the end of the sixteenth century, forks were quite unknown even at the court of the monarch. Neither at that period were they known in Sweden.

From the history of the travels of our countryman, Coryate, entitled "Crudities," first published in 1611, and afterwards in 1776, the author says he first saw them in Italy, and he was also the first person who used them in England. As his account of them is curious, we may be excused giving an extract, slightly altering the orthography.

"Here I will mention a thing that might have been spoken of before in discourse of the first Italian town. I observed a custom in all those Italian cities and towns through which I passed, that is not used in any other country I saw in my travels; neither do I think that any other nation in Christendom doth use it, but only Italy. The Italian, and also most strangers that are commorant in Italy, do always at their meals, use a little fork when they cut their meat. For while with their knife, which they hold in one hand, they cut the meat out of the dish, they fasten the fork, which they hold in their other hand, upon the same dish; so that whatsoever he be that, sitting in the company of any others at meals, should unadvisedly touch the dish of meat with his fingers, from which all at the table do cut, he will give occasion of offence unto the company, as having transgressed the laws of good manners, insomuch that for his error he shall be at least brow-beaten if not reprehended in words. This form of feeding I understand is generally used in all places of Italy;

their fork being for the most part made of iron or steel, and some of silver, but those are used only by gentlemen. The reason of this their curiosity is, because the Italian cannot by any means endure to have his dish touched with fingers—seeing all men's fingers are not alike clean. Hereupon I myself thought good to imitate the Italian fashion by this forked cutting of meat, not only while I was in Italy, but also in Germany, and oftentime in England, since I came home, being once equipped for that frequent using of my fork by a certain learned gentleman, a familiar friend of mine, one Mr. Lawrence Whitaker, who in his merry humour doubted not to call one at table *farsifer*, only for using a fork at feeding, but for no other cause."

In many parts of Spain, we understand that, *at present*, drinking-glasses, spoons, and forks are rarities. It is also said, that even in taverns in many countries, particularly in France, knives are not placed on the table, because it is expected that each person should have one of his own. This custom the modern French appear to have derived from their ancestors the ancient Gauls. But, as no person will eat any longer without forks, the landlords are obliged to furnish these, together with plates and spoons.

Among the Highlanders in Scotland, Dr. Johnson asserts, that knives have been introduced at table since the Revolution only. Before that period the men were accustomed to cut their meat with a knife they carry as a companion to their dirk. The men cut the meat into small morsels for the women, who used their fingers to put it into their mouths.

The use of forks at table was first considered as a superfluous luxury, and as such forbidden in convents, as appears from the records of the congregation of St. Maur.

MUSIC.

THE science of music, or rather of harmony, is extremely ancient—insomuch that, with respect to the latter, it is said to be coeval with Nature herself. But as it has relation to the science now in use, this, like most other arts, whose origin is very remote, is involved in obscurity; and in proportion to the astonishment and wonder excited by its uncommon powers, in a commensurate ratio does mystery, fable, and obscurity envelope its original. However, always remembering that it was from harmony,—

—"from heavenly harmony, this universal frame began."

Proceeding step by step, it had eventually attained in Greece a very early perfection. Collins, who is justly entitled to the distinguished station held by all pupils of nature and of the muses, who is peculiarly eminent for a just poetical spirit, thus speaks of the heavenly science in his Ode on the Passions—

"Arise, as in that elder time,
Warm, energetic, chaste, sublime;—
Thy wonders in that god-like age
Fill thy recording sisters' page.—
'Tis said, and I believe the tale,
Thy humblest reed could more prevail,
Had more of strength, diviner rage
Than all that charms this laggard age,
Even all at once together found
Cecilia's mingled world of sound."

It will be remembered, however, that the poet calculated as much upon the infant simplicity of nature as upon the uncommon powers of harmony; this consideration will certainly reconcile the apparent extravagance of the thought.

So great were the early powers of verse and harmony, that at one period the votaries of the muses were regarded as persons divinely inspired; they were the priests of man, his legislators, and his prophets. Insomuch was the possessor of the art, and the art itself reverenced, that the responses of the most eminent oracles were received in measured verse. Witness the response of the Delphian oracle received by the Athenian deputation, when Greece inquired for her wisest men, as given by Xenophon:—

"Wise is Sophocles, more wise Euripides,
But the wisest of all men is Socrates."

Music eventually claimed the most unlimited control over the affections of mankind, as could be proved by an infinity of instances; we shall mention one only from a well authenticated fact, and finely illustrated in that of Timotheus from "Alexander's Feast," by Dryden. We omit the hyperbolic representation of the raising of the walls of Thebes by the power of Amphion's lute, and the apparently incredible relations of the harmony of the harp of Orpheus, which are all personifications of natural effects, and which we have neither room, time, nor opportunity to explain in this place.

If its origin was as previously suggested by Collins, there is occasion to believe the shepherd's simple life afforded it first existence; in the native and wild notes of the pastoral reed, may be discovered the germ of a science as various as its effects are beautiful. We shall for the present presume the simple Pandean pipe was the first effort of the construction of musical instruments; its soft tone being analogous to the dulcet harmony of the voice.

We are led to suppose this from the evidence of ancient statuary, where those pipes are frequently discovered; and this will, perhaps, deduce its origin from the invention of the shepherd god, or oldest Pan. Nevertheless, the lyre, or harp, is alleged from records the most ancient, having at first but three strings, analogous to the three seasons of the primeval year; the treble typical of spring, the tenor resembling summer, and the bass representing winter.

The invention of that instrument, and of music altogether, is claimed in the pagan world by Amphion, a successor of Cadmus, the first king of Thebes, in Bœtia, who is reported, by the music of his harp or lyre to have built the walls of the city; Cadmus having erected the citadel only.

Flutes were first invented by Hyognis, the Phrygian, about the year 1506 before Christ, and first played on the flute the harmony, called Phrygian, and other tunes of the mother of the gods, of Dionysius, of Pan, and of the divinities of the country and the heroes. Terpander also, who was the son of Derdineus, the Lesbian, directed the flute players to reform the tunes of the ancients, and changed the old music, about the year 645 before Christ, as we are informed by the Parian Chronicle. The same Terpander, likewise, added three more strings to the lyre.

When Timotheus, the Spartan musician, was banished his native country for having increased his strings to the number of ten, he sought refuge at the court of Macedon, and accompanied his patron, Alexander, into Persia, when that prince conquered Darius.

From the sacred records of Judea, we may also infer the invention of musical instruments at a date long prior to either of the periods above mentioned, when they inform us in Genesis iv. 21, that Adah, one of the wives of Lamech, had two sons, the name of one of whom was

Jubal, who is said to have been "the father of all such who handle the harp and organ." This infers the anterior invention of that instrument.

Music consists of effects produced by the operation of certain sounds proceeding from the dulcet voice, or musical instruments, regulated by certain time, and a succession of harmonious notes, natural, grave, or flat, *i. e.*, half a note below its proper tone; and acute or sharp, *i. e.*, half a note above its proper key; and of such modulation of various tones, and of different value, and also of manifold denominations: the natural tones consisting of eight notes, with the addition of octaves, in various keys, with flats and sharps introduced to afford variety from the skill of the master, at different periods, to produce the most agreeable diversity in his composition; and sometimes according to the subject or words to which his music is adapted. Those musical notes, though proceeding from so small a number of radicals, are analogous to the incalculable, the endless forms, which orthography and rhetoric can afford to a well-informed orator, or elegant author, to embellish any subject. Thus from the definite number of twenty-four notes, varied in different degrees, by sharps, flats, semi-tones, &c., are produced all that is so magical, enthusiastic, and transporting in the empire of omnipotent music. Like as the alphabetic characters may be varied into myriads of forms suitable to every multifarious species of conversation or composition; in a word, a few musical notes in the hands of a master may be made by his skill to produce, from agreeable interchanges of time, harmony, &c., every variety of musical sentiment which can affect the human soul. A stronger proof cannot be adduced than will be found in the before-cited ode of "Alexander's Feast," by the truly poetic Dryden. In all which harmony and melody form conspicuous characteristics.

And of harmony, according to the learned Mr. Mason. The sense in which the ancient Greeks viewed harmony is as follows:—"They by that term understood the succession of simple sounds according to their scale, with respect to acuteness or gravity." Whilst it appears that by harmony, the moderns understand—"The succession of simple sounds, according to the laws of counterpoints." From the same authority—"By melody, the ancients understood the succession of simple sounds, according to the laws of rhythm and metre, or in other words, according to time, measure, or cadence. Whereas, the moderns understand by the same term what the ancients meant by harmony, rhythm and metre being excluded." "And the modern air is what the ancients understood by melody." Hence, from the preceding definitions, it appears that what is now called harmony was unknown to the ancients; and they viewed that term as we now see simple melody, when we speak of it as a thing distinguished from simple modulated air, and that their term, melody, was applied to what we now call air or song.

Should this be true, the long-contested difficulty, and that train of endless disputes, which has existed among the learned and scientific world so long, will instantly vanish. Should we suppose an ancient flute-player used an improper tone or semi-tone, or had he transgressed the mode or key in which he was playing, he committed an error in harmony; yet his melody might have been perfect, with respect to the laws of rhythm or metre; we should say of a modern musician, under similar circumstances, that he played wrong notes, or was out of tune, yet kept his *time*. Whoever made such a distinction would be allowed to possess a good ear for music, though the moderns would be inclined to call it an ear for melody or intonation. By the rules of musical conversation, we should be justified when we call an instrument out of tune inharmonious, although the intervals were nearly right.

By *harmonica*, the Greeks implied nothing more than that proportion of sound to sound, which mathematicians call *ratio*, or which would be understood in general musical conversation, by an agreeable succession of musical notes;—as ancient harmony consisted of the succession of simple sounds, so does modern harmony consist of the succession of chords.

Whether the *diatonic* scale be the effect of nature, or produced by art, has occasioned disputation between many; but without losing time or space, we are, we think, authorised, from general opinion, to observe, that compositions formed on it, and on the plan recommended by a lute organist, would produce sensations odiously disgusting to any musical ear.

The diatonic is the most simple genera in music, consisting of tones and major semi-tones; in the scale of which genus the smallest interval is a conjoint degree, which changes its name and place, that is, passing from one to another; a prominent air in this species of modern music is "God save the Queen," entirely diatonic, without modulation, by the intervention of a single flat or sharp.

It may not be unacceptable to our readers to add a few particulars of one of the greatest composers that ever existed; we allude to the eminently illustrious GEORGE FREDERICK HANDEL, a name dear to science, and entitled to the grateful veneration of every amateur in this divine art. He was born at Halle, in Upper Saxony, on the 24th of February 1684. Scarcely was he able to speak, before he articulated musical sounds. His father was a professor of the healing art as a surgeon and physician, then upwards of sixty, who intended his son for the study of the law. Grieved at the child's predeliction, he banished all musical instruments from his house. But the spark which nature had kindled in his bosom was not to be extinguished by the mistaken views of a blind parent. The child by some means or other contrived to get a little claverchord into a garret, where, applying himself after the family had retired to rest, he discovered means to produce both melody and harmony. Before he was seven years of age, the Duke of Weissenfells by accident discovered his genius, and prevailed on his father to cherish his inclination. He was accordingly placed with Zachan, organist of the cathedral of Halle; when, from nine to twelve years of age, he composed a church service every week. Losing his father whilst he was in that city, he thought he could best support his mother by repairing to Hamburgh, where he soon attracted general notice. This wonder of the age was then only fourteen, when he composed "Almeria," his first opera. Having quitted Hamburgh, he travelled for six years in Italy, where, at both Florence and Rome, he excited much attention: at both which places he produced new operatic performances. In that clime of the harmonious muse, he was introduced to, and cultivated the friendship of, Dominico, Scarlatti, Gaspurini, and Zotti, with other eminent scientific characters. He was particularly caressed and patronised by Cardinal Ottoboni, in whose circle he became acquainted with the elegant and natural Corelli. It was here he composed the sonata "Il trionfo del tempo," the original score of which is now in the Royal Collection. After which he went to Naples, where he set "Acis et Galatea," in Italian, to music. Returning to Germany, he was patronised by the Elector of Hanover, subsequently George the First. In 1710 he visited London, by permission of his patron, who had settled a pension of £200 per annum on him. In London he produced the opera of "Rinaldo," universally admired—equal with all his other productions that had preceded. He was compelled to leave, however reluctantly, the British shore, consistent with his engagement to his patron the Elector. He departed, not without exciting general regret, two years after his first arrival in this country. He soon appeared here again, however, and his return was welcomed like the rising of the genial orb of day before the wrapt Ignicolist! But now seduced by the favour which awaited him, he forgot to return. On the death of Queen Anne, who had also settled an annual pension of £200 upon him—equal to what he received from the Elector, his former patron—when that prince ascended the throne, Handel was afraid to appear before his majesty, till, by an ingenious contrivance of Baron Kilmarfyge, he was restored to favour, Queen Anne's bounty being doubled by the king; and the chief nobility accepted an academy of music under Handel's direction, which flourished for ten years, till an unfortunate quarrel occurred between him and Senesino, which dissolved the institution, and brought on a contest ruinous to the fortune and the health of our musician.

He was particularly patronised by the Earl of Burlington, the Duke of Chandos, and most of the distinguished nobility of Great Britain.

Having restored his health at the baths of Aix-la-Chapelle, he for the future chose sacred subjects, which were performed at his theatre in Lincoln's Inn Fields, Covent Garden, and Westminster Abbey. He died in April, 1759, aged seventy-five, and was buried in Westminster Abbey, where he was honoured with a public funeral, six peers supporting the pall; the very reverend and truly learned translator of "Longimus," Dr. Pearce, the Dean, and then Bishop of Rochester, performed the funeral service with a full choir.

He had been a great benefactor to numerous public charities. The funds of the Foundling Hospital were improved through him with the amazing sum of £10,299. The organ in its chapel, and the MS. score of his "Messiah," were a present and a donation to the foundation from him. He left an amiable private as well as a good public character behind him.

His character as a composer is too well appreciated by the British public to require any remarks from our feeble and inharmonious pen.

SEALING-WAX, SEALS, &c.

BESIDES metals, five other mediums are enumerated by ancient writers, wherewith letters and public acts were sealed, viz., *terra sigillaris*, cement, paste, common wax and sealing-wax. That the terra sigillaris was used by the Egyptians, we have the evidence of Herodotus, and which, by inference, is strengthened by that of Moses who speaks of seal-rings or signets, whence we may safely infer, that they had a medium of some sort, wherewith they sealed. This *lacuna* Herodotus supplies, affirming it in direct terms, and assigning a name to the substance they used for that purpose.

This circumstance was only rendered questionable by Pliny, who alleges the Egyptians did not use those things.

Herodotus thus expresses himself: "The Egyptian priest bound to the horns of cattle fit for sacrifice pieces of papyrus with sealing-earth, on which they made an impression with the seal; and such cattle could only be offered up as victims."

Lucian speaks of a fortune-teller who ordered those who came to consult him, to write down on a bit of paper the questions they wished to ask, to fold it up, and seal it with clay, or any other substance of a like kind.

Such earth appears to have been employed in sealing, by the Byzantyne emperors; for we are told that, at the second Nicene Council, image worship was defended by one saying, "No one believed that those who received written orders from the Emperor, and venerated the seal, worshipped on that account the sealing-earth, the paper, or the lead."

Cicero relates that Verres, having seen in the hands of his servants a letter written to his son from Agrimentum, and observing on it an impression in sealing-earth, he was so pleased with it that he caused the seal-ring with which it was made to be taken from the possessor.

Also, the same orator, in his defence of Flaccus, produced an attestation sent from Asia, and proved its authenticity by its being sealed with Asiatic sealing-earth; with which, he told the judges, all public and private letters in Asia were sealed: and he showed on the other hand, that the testimony brought by the accuser was false, because it was sealed with *wax*, and for that reason could not have come from Asia. The scholiast Servius relates, that a sybil received a promise from Apollo, that she should live as long as she did not see the earth of the island of the Erythræa, where she resided; that she therefore quitted the place, and retired to Cumae, where she became old and decrepid; but that having received a letter sealed with Erythræn earth, when she saw the seal, she instantly expired.

No one, however, will suppose that this earth was used without preparation, as was that to which is given the name of *creta* chalk; for, if it was of a natural kind, it must have been of that kind called *potter's clay*, as that clay is susceptible of receiving an impression, and of retaining it subsequent to hardening by drying. It is believed that the Romans, under the indefinite term *creta*, often understood to be a kind of potter's earth, which can be proved by many passages in their numerous writers. Columella speaks of a species of chalk of which wine-jars and dishes were made, of which kind it is conjectured Virgil speaks when he calls it adhesive. The ancient writers on agriculture give precisely the same name to marl, which was employed to manure land: now, both chalk and marl, in their natural state, are extremely inapplicable to the purpose for which we are led to believe the *terra sigillaris* was used; therefore, admitting the Roman *creta* was composed of them, those substances must naturally have undergone some laborious process, in order to render them proper for the purpose to which they were applied.

Notwithstanding none can feel a higher respect for Professor Beckmann, to whom we are indebted for many of the preceding observations, than we do, yet strongly as we are influenced with this impression, we cannot help observing, consistent with that duty we owe to the public, that we cannot divest ourselves of the opinion that he is only trifling with the public feeling, perhaps for the ostentatious display of his own learning: so many objections of so little weight are raised, that he really appears to write for the purpose of raising new objections to passages, which, in our comprehension, are extremely simple. We cannot help applying to him a passage which occurs in a song of the Swan of Twickenham, who sings:

"Gnawed his pen, then dashed it on the ground,
Striking from thought to thought, a vast profound.
Plunged for the sense, but found no bottom there,
Yet wrote, and floundered on in mere despair."

We would not be illiberal or capricious, nor do we presume to any extra portion of intelligence; yet, we think we can in a few words discuss the topic, and perhaps, satisfactorily, on which he has employed so many pages. Those terms which have troubled the professor with learned difficulties really appear to us susceptible of an easy interpretation, and applicable to both or either of the senses in which they are used, as are any words in the language of ancient Rome. Accordingly, we find the term *creta* implies either chalk, fuller's clay, loam, white paint, or Asiatic earth, termed creta Asiatica; and, in brief it appears a mere generic name for any kind of earth, raised from below the surface of the soil: this is its true sense. But there cannot be a question, from what is known of the preparation of clay and earth for *terra cotta* and other plastic purposes, which undergo a variety of washings, kneadings, &c., that similar preparations were requisite, in order to bring it to so curious, so delicate a purpose as that to which the terra sagillaris was applied. And *fosse*, in the sense used by Varre, admits of nearly a similar description, it appearing as a pronomen for the same thing; and indicates either peat, marl, loam, chalk, or any earthy substance which may be raised from below the terrestrial surface.

We have evidence every day in our fruit shops, that in certain countries this kind of earth is yet employed for closing up jars of dried fruits brought from Oporto, Smyrna, and other countries; as these appear to be composed of white chalk of a texture somewhat similar to common mortar. The warmth of the atmosphere, where it is used, soon hardens and prevents the passage of air to the contents; the jars themselves being oftentimes only dried in the sun.

Thus it appears that prepared earths were first used for the purpose of sealing; their adhesive, or, as Virgil has it, their tenacious qualities, being wonderfully improved for manual labour. Next, paste was employed, prepared from dough.

To paste succeeded common wax, sometimes slightly tinctured with a green tint, the effect of endeavouring to give it a blue colour, as vegetable blues turn green by the process of heat employed in melting; whilst mineral or earthy blues all sink to the bottom, from superior gravity.

This was the material employed in sealing public acts in England, as early as the fifteenth century. We have an anecdote of the Duke of Lancaster having no seal to ratify a deed between him and the Duke of Burgoyne, but from what appears in the attestation, which, with the instrument itself, according to the general custom of the day, runs in rhyme thus:

"I, John of Gaunt,
Doe gyve and do graunt,
To John of Burgoyne
And the heire of his loyne
Sutton and Putton
Untill the world's rotten."

The attestation runs thus:

"There being no seal within the roof,
In sooth, I seal it wyth my tooth."

A good example is this of the simple brevity of the time, and a severe lecture upon the eternal repetitions of our modern lawyers, whereby the limitations and special uses of deeds are made, perhaps, not according to the necessities of the case, but are lengthened from selfish purposes.

The Great Charter, which gives an assurance of the rights of Englishmen, is sealed with white wax; as may be seen in the British Museum.

The first arms used as a seal in England, were those of the tyrannical subjugator of English rights, William, commonly called the Conqueror, and they were brought from Normandy.

Although Fenn, in his collection of original Letters of the last half of the fifteenth century, published in London,

1787, has given the size and shape of the seals, he does not apprise us of what substance they were composed. Respecting a letter of 1455, he says only, that "the seal is of red wax," by which, it is presumed, he means common wax; and though, perhaps not equal in quality to such as is now used, yet it was made of nearly similar materials. Tavernier, in his Travels, says, that in Surat gum-lac is melted and formed into sticks, like sealing-wax. Wecker also gives directions to make an impression with calcined gypsum and a solution of gum or isinglass. Porta, likewise, knew that this might be done, and, perhaps, to greater perfection with amalgam of quicksilver.

Among the records of the Landgrave of Hesse-Cassel, are some letters of 1563, sealed with red and black wax. In the family of the Rhingrave, Philip Francis von Daun, the oldest letter sealed with wax, known in Germany, is found, of the date of August 3, 1554; it was written from London, by an agent of that family, of the name of Gerrard Herman. The colour of the wax is dark red, and very shining.

The oldest recipe known in Germany for making sealing-wax, was found by M. Von Murr, in a work by Samuel Zimmerman, citizen of Augsburg, published in 1759. The copy in the library of the university of Gottingen is signed by the author himself.—"To make hard sealing-wax, called Spanish wax, with which, if letters be sealed, they cannot be opened without breaking the seal; take beautiful clear resin, the whitest you can procure, and melt it over a slow coal fire. When it is properly melted, take it from the fire, and for every pound of resin, add two ounces of cinnabar, pounded very fine, stirring it about. Then let the whole cool, or pour it into cold water. Thus you will have beautiful red sealing-wax.

"If you are desirous of having black wax, add lamp-black to it. With smalt or azure, you may make blue: with white-lead, white; and with orpiment, yellow.

"If, instead of resin, you melt purified turpentine in a glass vessel, and give it any colour you choose, you will have a harder kind of sealing-wax, and not so brittle as the former."

It may be remarked, that in these recipes for the fabrication of sealing-wax there is no mention of gum-lac, which is known at present as a chief ingredient in the composition of this article.

Zimmerman's sealing-wax approaches very near to the quality of that known as *maltha*, whence we may conclude, that the manufacture of it did not originally come from the East Indies. The most ancient mention of sealing-wax occurs in a botanical work, treating of the history of aromatics and simples, by Garcia ab Horto, published at Antwerp in 1563, where the author, speaking of gum-lac says, that those sticks used for sealing letters are made of it; at which time sealing-wax was common among the Portuguese, and has since been manufactured chiefly in Holland.

M. Spiess, principal keeper of the Records at Plessenberg, says, respecting the antiquity of *Wafers*, in Germany, that the most ancient use of them he has known, occurs in a letter written by D. Krapf, at Spires, in 1624, to the government of Bayreuth.—The same authority informs us that some years after, the Brandenburg factor at Nuremberg sent such wafers to a bailiff, at Osternohe. During the whole of the seventeenth century, wafers were not used in the Chancery at Brandenburg, and only by private persons there.

Seals, it appears, from certain passages of Egyptian history, parallel with, and perhaps anterior to the Israelitish ingress, were formed or cut in emeralds, the native produce of that country. Other precious stones, metals, steel, lead, and a variety of materials, but chiefly of a hard and precious kind, have been always employed for that purpose.

BLACK-LEAD PENCILS.

THE period when this semi-metallic substance was introduced, for the purpose for which it is now applied, cannot with certainty be ascertained, as no record is found of the transaction: by the common expedient of inference, however, we certainly may conclude, it was in very remote ages; for transcribers of MSS. upwards of one thousand years ago, used a substance somewhat resembling it in effect.

But, perhaps, the antiquity of the use of black-lead pencils cannot be so well determined from diplomatiques, as their frequency might be proved from mineralogical writers. The first mention of this discovery occurs in the works of Gesner, who, in his "Book of Fossils," published in 1565, says that the British people had pencils for writing, with wooden-handles inclosing a piece of lead, which he believed to be an artificial composition; and it was called *stimmi Anglicanum*; which seems to import that it was a British production; and we should consider, from the name of British antimony being given to it, that it might have been Cumberland black-lead.

About thirty years afterwards, Cæsalpinus gave a more perfect account of it:—he says it was a lead-coloured, shining stone, as smooth as glass, and appeared as if rubbed over with oil; it gave to the fingers an ash-grey tint, with a plumbeous brightness; and, he adds, pointed pencils were made of it, for the use of painters and draughtsmen. A closer description of the substance than this cannot be discovered.

Somewhere about three years afterwards, a still more perfect description was furnished by Imperatis; who says,

"It is much more convenient for drawing than pen and ink, because the marks made with it appear distinct upon a white ground, also, in consequence of its brightness, show themselves on black, and can be preserved or rubbed out at pleasure. This mineral is smooth, appears greasy to the touch, and has a leaden-colour, which it communicates with a metallic brightness. It can resist, for a long time, the strongest fire, and even from it requires more hardness; it has, in consequence, been thought to be a species of *talc*. This, in the arts to which it is applied, is a property which greatly enhanceth its value, being manufactured into crucibles, &c., with clay. These vessels are capable of enduring the strongest heat of a chemical furnace."

Sometimes this lead is foliaceous, and may be crumbled into small pieces or scales; but frequently found denser and more strong. This latter is what writing pencils should be made of; but the former being more frequently found, and, also, coming from the refuse of the workmen, is too often mixed up with some glutinous substance, and there is every reason to suppose it to be enclosed in the groove in a plastic state; these pencils are commonly hawked about our streets by pedlars and Jews; of purchasing which people should be cautious, as they are, in general, utterly worthless.

Robinson, in his Essay towards a natural History of Westmoreland and Cumberland states, that, at first, the country people round Keswick marked their sheep with black-lead. Afterwards, they discovered the art of employing it in their earthenware, and also to preserve iron from rust. The same writer says, the Dutch use it in dyeing, to render black more durable; and that they buy it in large quantities for that purpose. But their application of it for dyeing, we should consider as highly questionable.

The mode of eradicating black-lead by means of an elastic gum, called caoutchouc, or, Indian-rubber, was, we have been informed, first discovered in England somewhere about sixty years ago.

COLOURED GLASS.

THE manufacture of glass we find was quite common in Ethiopia, Syria, Assyria, and other Eastern countries, in the earliest ages of the world, as Diodorus Siculus informs us, who says, the Ethiopians enclosed in glass, the bodies of their parents and friends; we doubt, however, that on this point, the historian was deceived. But it really appears probable that soon after the art of making glass was discovered, the idea of communicating to it some colours would easily present itself. This probability appears increased, when it is recollected that much care is requisite to render glass perfectly colourless. As the various metallic particles with which stone and sand abound, (these being the chief ingredients of which glass is composed, and which gradually give tints in fusion,) will almost unavoidably communicate some hue or other, therefore the perfection of glass is to have it perfectly colourless.

But with respect to coloured glass; so frequently have people been imposed upon by having coloured glass sold to them for valuable stones, that some conscientious authors have very laudably and carefully abstained from lending the benefit of instruction in its manufacture, by publishing the method.

The Egyptian artists were so famous in the manufacture of glass, that the Romans were content to receive this article from the glass-houses in Alexandria, and did not interfere in endeavouring to procure the art themselves, until the latter part of the empire.

We read that an Egyptian priest made a present to the Emperor Adrian of several beautiful glass cups, which sparkled with many colours; and such value did that august personage place upon these toys, that he ordered them to be used only on high feasts and solemnities.

Strabo relates, that a glass manufacturer of Alexandria informed him that an earth was found in Egypt, without which the valuable coloured glass could not be made. It has been thought by some, the glass earth here meant was a mineral alkali which was readily found in Egypt, serving to make glass; but this author speaking expressly of coloured glass, it has been suggested as probable, the alkali above named could not have reference to what the artisan intended to imply, but that it must be referred to some metallic earth or manganese.

One Democritus is named by Seneca, as having discovered an art of making artificial emeralds; but it has been conjectured that what the philosopher meant was the art of communicating colour to natural rock crystal, or colouring glass already made, so as to resemble stones, which is a process performed by cementation. Directions have been furnished for this purpose by Porta, Neri, and others; but it is discovered that the articles so coloured are liable to such accidents in the process, that it is next to impossible to render things of any size tolerably perfect, so as to bear cutting afterwards.

In the Museum Victorium at Rome, there are shown a chrysolite and an emerald, both perfectly well executed, and thoroughly transparent, without a blemish.

We have not from the ancients an account of what process they employed; but it must be evident that nothing less than metallic calces could have been used; and for this evident reason, that any other substance could not have resisted the influence of the necessary heat. The last century has, however, produced certain artists in northern European nations, who have adopted a method of employing the precious metals, to communicate a tincture to glass in the process of making, where iron, &c. were originally only used; and their endeavours have been attended with singular success.

By means of an amalgam of gold, or a solution in *aqua regia*, and precipitated with a solution of tin, the metal then assuming the appearance of a rich purple coloured powder; so prepared, it is mixed with the best *frit*, and then called the precipitate or gold calx of Cassius, the inventor of gold purple, or mineral purple.

This precipitate communicates a rich ruby coloured purple, so perfect that it is impossible to discover the deception, without the substances be tried by the usual means—cut with a diamond or a prepared file.

We have had in England some very eminent artists in the practice of staining glass, and also for making artificial representations of various precious stones.

Although the professed object of alchemy has now met with that contempt it merited—because, notwithstanding the immense sums which have been expended, the time lost, and unprofitable labour employed in the unavailing search after what probably never will be found—yet the labour lost and money expended has not been totally useless, since it has served to open the seals which secured chemical science to the modern world; and which is the chief, if not the sole advantage it can claim over antiquity for superiority of information.

Painting on glass, but, perhaps, staining had been a more appropriate expression, or, properly speaking, in enamel, with the preparations for colouring in mosaic work, may, to a certain extent, be justly considered as branches of the art of colouring glass; in all which there is no colour more difficult to be attained than a beautiful red; it now is, and ever has been, most difficult, consequently the dearest colour. The presumed ignorance of ancient artists in preparing this colour has afforded some reason, it is said, to suppose the ancients knew of no other substance proper for that purpose but calx of iron, or manganese. To this we may reply, many specimens are found which show they were not so ignorant in that art, and that it is more than probable the same jealousy which is found to exist in modern days among artizans might prevent our sagacious predecessors from publishing the secrets of their respective professions to the world. We contend, that as the materials must then have had existence, which have been since so successfully employed, pray what was the reason the ancients should not avail themselves of their benefit? In all the higher speculations of science and arts, where the great and superior energies of genius were requisite, this perfection in the ancients far surpassed any exertions which have been since achieved by the moderns. To instance one artist and one art solely, we name the great Praxiteles, so famous in the art of statuary, whose works were a model of perfection.

ETCHING ON GLASS AND GLASS CUTTING.

WITHOUT entering into the history of the lapidary's art, we only propose to speak of those things which ancient and modern authors have said upon the art of engraving on glass, observing, that it was an art anciently known to both the Greeks and Romans; although it appears extremely probable, that from their expressed ignorance of many of those properties which modern chemistry has discovered to belong to matter, they were ignorant of the art of etching on glass.

From antique specimens still preserved, a doubt cannot for a moment be suffered to exist on our minds, but that the art of engraving upon glass was familiar to the Greek artists, who formed upon glass both linear figures, and in relievo, by the same means as are now employed for nearly the same purpose, if we can place any confidence in an able and learned lapidary, Natter, who has established, that the ancients employed the same kind of instruments for this purpose, or nearly such as are now in use; abating, perhaps the use of diamonds, and the dust of that precious material, for which it is conceived they used emery powder, and the dust of glass.

From what is related by Pliny, it certainly appears that they used the lapidary's wheel, an instrument moving in a horizontal direction over the work-table.

Some have thought that drinking cups and vessels may have been formed from the glass whilst in a state of fusion, by means of this wheel; to this they think those words of Martial refer, where he says, *calices audaces*, having reference to the boldness of the artisan's touch; those vessels he was constructing often broke under the last touch he bestowed upon his transparent labours, although, perhaps, of costly value; these accidents must of necessity have rendered those articles extremely expensive.

There are not wanting many who affirm the art of glass-cutting, with the instruments necessary for that operation, to be of modern invention. Those assign it to the ingenuity of Caspar Lehmann, originally an engraver on iron and steel, and who, as Beckmann informs us, made an attempt, which succeeded, in cutting crystal, and afterwards glass in the same manner. This artist, we are told, was in the service of Rodolphus, the second emperor of that name, who, in the year 1609, besides giving him valuable presents, conferred on him the title of lapidary and glass-cutter to his court, and gave him a patent, allowing him the exclusive privilege of exercising this new art. He worked at Prague, where he had an assistant of the name of Zacharias Belzer; but George Schwanhard, one of his pupils, carried on the business to a much larger extent. The last named was a son of Hans Schwanhard, a joiner at Rothenburg, and was born in 1601; at the age of seventeen he went to Prague, to learn the art of cutting glass from Lehmann. His good behaviour won so much upon the affections of his master, that on his death in the year 1622, he left him his heir. Schwanhard succeeded in obtaining a continuation of the patent from the emperor, and removed to Nuremburg, where he wrought for many of the nobility of that district. This was, we believe, the occasion of that city claiming the honour of being the birth-place of this new art. In the year 1652, he worked at Prague, and also at Ratisbon, by command of the Emperor Ferdinand III.; and he died in 1676. He left two sons, who both followed the lucrative employment of their father. Afterwards Nuremburg produced many expert masters in the art, who, from the improvement in the tools, and also from discovering more economical modes of using them, were enabled to execute the orders of the public at a more moderate rate than had been previously charged for some articles. Those latter masters likewise brought this art to a much greater degree of perfection. Notwithstanding Zahn was of the same country, and must have been apprised of the facts previously stated, yet he mentions it as a very recent invention at Nuremberg, at the time he published his "Oculus Artificial." He also furnishes a plate, giving at the same time a description of the various instruments employed. However, that this invention is not purely *novel*, may be perceived from those facts we have already submitted.

It should be stated that before this latter re-introduction, artists used, with a diamond, to cut figures upon glass in almost every form, as far as the representation by lines went. The history of diamonds has been presented to the public by Mr. Mawe, in his observations on the diamond districts of Brazil. It appears to be yet undetermined whether the ancients used that stone for the purpose of cutting others; upon this point Pliny appears to be satisfied that they did.

Solinus and Isidore both express themselves in a manner the reverse. But although this may leave us in some doubt, it appears pretty clear that they did not attempt to cut that valuable production with its own dust, or to give it different faces, or render it more brilliant by the same means. If this point was settled, there could be no great difficulty in affirming or negativing the fact of their engraving upon that stone. Thus doubts appear to increase on this head, for Mariette denies that they did; Natter appears uncertain; and Klotz asserts with confidence it was certain. His authority, to be sure, has been considered not to be of much weight.

The proper question, however, appears to be, whether the Greeks and Romans used diamonds for cutting and engraving other stones or glass. Natter, in his work already noticed, thinks they were employed on some antique engravings. His authority is deserving respect. But if they were employed on other stones, the authority which at present directs us, confidently alleges they did not employ them in cutting glass; but he points out the mode in which that article was wont to be divided, in the following terms: "They used for that purpose emery, sharp-pointed instruments of the hardest steel, and a red-hot iron, by which they directed the rents at their pleasure."

The first mention which appears to occur of the use of the diamond for this purpose, is recorded of Francis I. of France, who, fond of the arts, sciences, and new inventions, wrote a couple of lines with a diamond, on a pane of glass in the Castle of Chambord, to let Anne de Pisseleu, Duchess of Estampes, know that he was jealous.

About 1652, festoons and other ornaments, cut with a diamond, were made on Venetian glasses; then considered the best. Schwanhard was a professed adept in that art; and since his time an artist of the name of John Rost, of Augsburg, cut some drinking glasses, which were purchased by the Emperor Charles VI., at an extravagant price.

ETCHING ON GLASS.

An acid to dissolve siliceous earth was discovered as late as 1771, by the celebrated chemist Scheele, in *sparry fluor*. It is conceived that this cannot be of older date than that period; but it is alleged that an acid was discovered as early as the year 1670, by Henry Schwanhard. It being said that some aquafortis had dropped, by accident upon his spectacles, the glass being corroded by it, he thence learned to improve the liquid that he could etch figures and write upon glass. How he prepared this liquid is a secret which has not been revealed. The *Teutsche Akedemie* says on this subject, that he, by the acuteness of his genius, proved that which had been considered impossible could be accomplished; and found out a corrosive so powerful that the hardest crystal glass, which had hitherto withstood the force of the strongest spirits, was obliged to yield to it, as well as metals and stones. By these means he delineated and etched, on glass, figures of men, in various situations, animals and plants, in a manner perfectly natural, and brought them to the highest perfection.

The glass proposed to be etched is made perfectly clean and free from grease; then the figure is covered with a varnish; then an edge of wax being raised round the glass, the acid is poured in, and the whole ground on the exterior of the figures appears rough, whilst the figure is preserved in its original beauty of outline, bright and smooth. This is the mode the inventor adopted.

Professor Beckmann says, he mentioned this ancient method of etching upon glass, to an artist of the name of Klindworth, who possessed great dexterity in such arts, and requested him to try it; he drew a tree with oil varnish and colours on a plate of glass, applied the acid on the plate in the usual manner; after it had been upon the plate for a sufficient time, poured off, and the plate afterwards cleaned of the varnish, a beautiful tree was left bright and smooth, with a rough back-ground. It is conceived that many great improvements may yet be made in this process.

It appears that no other acid than that produced by the sparry fluor is capable of corroding every kind of glass, though Baume, in his "Chemique Experimentale," says, that many kinds of glass may be corroded by the marine and vitriolic acids.

In this state of uncertainty was the public mind till the year 1725, when it was thought that a recipe, older than that previously mentioned, might possibly be discovered. Accordingly, in that year, in the month of January, the following is said to have been transmitted to the publisher of the "Œkonomische Encyclopedie," by Dr. John George Weygand, of Goldingen, which is reported to have belonged to Dr. Matthew Pauli, of Dresden, then deceased; with which the last named gentleman had etched, on glass, arms, landscapes, and figures of various kinds. We find, that in it, very strong acid of nitre was used, which entirely disengages the acid of sparry fluor, though the vitriolic acid has been commonly employed, and figures thus produced will appear as if raised above the plane of the glass.

This sparry fluor is found abundantly in Derbyshire, as well as in the mines of Germany. Theophrastus is the first who notices the effect of sparry fluor, by observing that there are certain stones which, when added to silver, copper, and iron ores, become fluid. It appears that Cronstedt was the first systematic writer who gave it a name.

When *spiritus nitri per distillationem* has passed into the recipient, ply it with a strong fire, and when well dephlegmated, pour it, (as it corrodes ordinary glass,) into a Waldenburg flask; then throw into it a pulverised green Bohemian emerald, otherwise called *hesphorus*, (which, when reduced to powder and heated, emits in the dark a green light,) and place it in warm sand for twenty-four hours. Take a piece of glass, well cleaned, and freed from all grease by means of a ley; put a border of wax round it, about an inch in height, and cover it equally all over with the above acid. The longer you let it stand, so much the better; and at the end of some time the glass will be corroded, and the figures which have been traced out with sulphur and oil varnish will appear as if raised above the plane of the glass.

HYDROMETERS.

THE Hydrometer is an instrument for admeasuring liquids; by it the strength or specific gravity of different fluids is discovered, by the depth to which it sinks in them. It has been chiefly used for discovering the contents of different salt waters, without analysis, and is now almost entirely used by persons connected with the spirit-trade, to ascertain the different degrees of strength, and what alloy they will bear; hence its utility to the manufacturer and the excise-officer is apparent.

The laws respecting the comparative weight of different fluids, as well as of solid bodies immersed in them, was first discovered by that great geometrician Archimedes. It may be far from improbable that Archimedes constructed that instrument himself; and if it should appear that he did, it must have happened two hundred and twelve years before the Christian era.

The most ancient mention of this instrument by its specific name, occurs in the fifth century of our era, upon the following occasion. The anecdote is very singular and affecting, and also evinces the incapacity of humanity to act consistent and as it ought, when we suffer ourselves to be directed by passions unworthy of the human character.

It is first discovered in those letters of Synesius to the philosophic and beautiful Hypatia. We trust we may be excused the liberty we propose to take in detailing this circumstance, which is comparatively little known; and as its interest also recommends it, this furnishes an additional motive.

Hypatia was the daughter of Theon, an eminent mathematician of Alexandria, some of whose writings are still extant. By her father she was instructed in the mathematics, and from other great men, who at that period abounded in Alexandria, she learned the Platonic and Aristotelian philosophy, and acquired such a knowledge of these sciences, that she taught them publicly, with the greatest applause. She was young and beautiful, had a personable figure, was sprightly and agreeable in conversation, though, at the same time, modest; and she possessed the most rigid virtue, which was proof against every temptation. She conducted herself with so much propriety towards her lovers, that they never could obtain more than the pleasure of her company, and hearing her discourse; and with this, which they considered as an honour, they were contented. Those who were so daring as to desire further communion she dismissed; and even destroyed the appetite of one of her admirers, who would not suffer her to philosophise, by means of some strong preparation, which others appear not to have since imitated.

She suffered so cruel a death, that had she been a Christian, and suffered from Pagan error, her name would have been ranked among its most honoured victims in the list of martyrology; but being a Pagan, and suffering from the persecution of superstitious and anti-Christian zeal, she is honoured among the foremost of martyrs to celestial philosophy.

The name of the Christian patriarch, at that period in Alexandria, was Cyrill, whose family had, for upwards of a hundred years before his time, produced bishops, who had been much more serviceable to their own family connections than they had ever been towards the propagation of the Christian faith. The present was proud, litigious, and revengeful, vindictive and intolerant to the last degree; his ignorance debasing his own character as a man, and scandalising the religion of which he was so unworthy a minister. He stupidly conceived himself sanctioned in everything which his foolish and mistaken ideas might dictate to be for the glory of God, and acted as a persecutor, prosecutor, judge, and executioner: he had condemned Nestorias without hearing his defence. As the city of Alexandria was then very flourishing on account of its extended commerce, the emperor had there allowed greater toleration and more peculiar privileges to all religions, than in any other place: it consequently contained, among others, a great number of Jews, who carried on a most extensive trade, as well as a great many Pagan families. In the eyes of the bigot Cyrill this was wrong; he would have the sheep-fold clean, and the Jews must be banished. The governor, however, who was a man of prudence and sober discretion, much better acquainted with the real interests of the city, opposed a measure he saw replete with mischief, and even caused to be condemned to death a Christian profligate, who had injured the Jews. This malefactor was, by the express order of Cyrill, buried in the church as a martyr; and he collected an army of five hundred lazy monks, who abused the governor in the public streets, and excited an insurrection among the people against the Jews, so that the debased race of Abraham was expelled from the city where they had so long existed unmolested from the time of Alexander the Great.

Cyrill, one day, whilst looking for objects of persecution, saw a number of carriages, attended with servants, belonging to the first families in the city, before a certain house. Inquiring what was the cause of the assembly, he was informed that it was the habitation of the lovely Hypatia, who, on account of her extensive learning and very eminent talents, was visited by people of the first respectability. This afforded to the malignant priest a sufficient object for the exercise of his jealousy against the meritorious, the unoffending, the beautiful Hypatia. He from that moment resolved upon her destruction. Accordingly he lost no time in exciting his myrmidons, the monks and priests, those who should have been the ministers of that religion which they professed to teach, to destroy the fair philosopher. They accordingly, with diabolical rage, and instigated by infernal cruelty, took the earliest opportunity to seize her, hurried her to the church—the temple of peace and good-will—which they violated by an offence at which humanity must shudder; having torn the clothes from her delicate form, they tore the flesh from her bones with potsherds, then dragged her mangled body about the city, and afterwards burnt it.

This demoniacal tragedy took place in the year 415, and was perpetrated by the professed servants of Him who came into world to save those which were lost—to preach peace and good-will to all men. The impressions which such an event made upon people of every persuasion may be conceived; they admit not of description from a feeble pen: but we may ask the question, was it such a transaction as was calculated to make converts to the doctrines of Christianity?—whose avowed motive and maxim is, in the words of Milton,

"By winning words, to conquer willing hearts,
And make persuasion do the work of fear."

All historians are not agreed in some circumstances of the preceding relation; but they generally unite in bestowing praise upon Hypatia, whose memory was long honoured by her grateful and affectionate scholars, among whom was Synesius, of a noble Pagan family, who had cultivated philosophy and the mathematics with the utmost ardour, and who had been one of her most intimate friends and followers. On account of his learning and virtues, many eminent talents, and open disposition, the inhabitants of Ptolemais were desirous he should be bishop, having been previously employed on many public and important concerns with success. After modestly desiring, for a long period, that they would fix their choice upon a more worthy object, they still persisting, he assented, upon condition that he was not to believe in the resurrection, to which he could not at that time bring his internal conviction: he suffered himself to be baptised, and became their bishop; he was confirmed by the orthodox patriarch Theophilus, the predecessor of Cyrill, to whose jurisdiction Ptolemais belonged: he afterwards renounced his error respecting the resurrection. This learned man evinced his gratitude to Hypatia, by the honourable mention which he made of her in some of his writings, still preserved.

In his fifteenth letter to her, he tells Hypatia, that he was so unfortunate, or found himself so ill, that he wished to use an hydroscopium (the Greek for hydrometer), and he requests that she would cause one to be constructed for him. He says, "It is a cylindrical tube, of the size of a reed or pipe; a line is drawn upon it lengthways, which is intersected by others, and these point out the weight of water. At the end of the tube is a cone, the base of which is joined to that of the tube, so that they have both only one base. This part of the instrument is called *baryllion*. If it be placed in water, it remains in a perpendicular direction, so that one can readily discover by it the weight of the fluid."

Petau, who published the works of Synesius, in the year 1640, acknowledges that he did not understand this passage. An old scoliast, he says, who had added some illegible words, thought it was a water-clock; but the ellepsydra was not immersed in water, but filled with it.

He therefore thought that it might allude to the chorobates, which Vitruvius describes as an instrument employed in levelling; but it appears that Synesius, who complained of ill health, could have no occasion for such an instrument. Besides, no part of that instrument he describes, has any resemblance to the one described by Synesius.

From the works of Fermat, an excellent mathematician, and a very learned man, well acquainted with antiquities and the works of the ancients, we give the following explanation concerning the hydroscopium of Archimedes, as this article would be incomplete without it:—

"It is impossible," says he, "that the *hydroscopium* could be the level or *chorobates* of Vitruvius, for the lines on the latter were perpendicular to the horizon, whereas the lines on the former were parallel to it. The hydroscopium was undoubtedly a hydrometer of the simplest construction. The tube may be made of copper, and open at the top; but at the other end, which, when used, is the lowest, it must terminate with a cone, the base of which is added to that of the tube. Lengthwise, along the tube, are drawn two lines, which are intersected by others, and the more numerous these divisions are, the instrument will be so much the more correct.—When placed in water it sinks to a certain depth, which will be marked by the cross-lines, and which will be greater, according to the lightness of the water." A figure which is added, might have been dispensed with. When a common friend of Fermat and Petau showed it to the latter, he considered it to be so just, and explanatory of the real meaning of Synesius, that he wished to be allowed the opportunity of introducing it in a new edition of the works of Synesius.

FINIS.

www.ingramcontent.com/pod-product-compliance
Lightning Source LLC
Chambersburg PA
CBHW071414180526
45170CB00001B/96